DOWNWIND FASTER
THAN THE WIND

Sailing explained by
Newtonian physics
and Galilean relativity

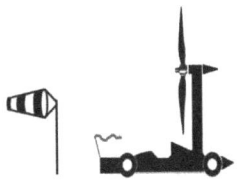

DOWNWIND FASTER THAN THE WIND

Sailing explained by Newtonian physics and Galilean relativity

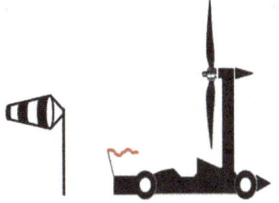

BY NICHOLAS LANDELL-MILLS

ABOUT THE AUTHOR

THE AUTHOR FIRST developed the ideas presented in this book as a child, while sailing Loch Longs with his grandfather in Aldeburgh, Suffolk, UK. The author is British and was born in 1966 in Botswana. He is a graduate of the University of Edinburgh, Edinburgh, UK where he was awarded a M.A. degree. He previously worked in finance for twenty-five years and is now an independent researcher in applied physics.

Published by Boatswain Books 2021

Copyright © Nicholas Landell-Mills 2021

The full range of books published by Boatswain Books can be found at: www.boatswainbooks.uk

Nicholas Landell-Mills has asserted his right under the Copyright, Designs and Patents Act, 1988, to be identified as the author of this work.

First published in Great Britain 2021

A CIP catalogue record for this book is available from the British Library.

ISBN: 978-1-912724-24-6

All rights reserved. No part of this publication may be reproduced, stored in a retrieval system or transmitted in any form by any means, electronic, mechanical, photocopying, recording or otherwise without the prior written permission of the publisher and the copyright holders.

ABSTRACT

NEW ANALYSIS SHOWS how Newton and Galileo provide a straightforward and useful explanation of sailing.

According to Newton, the sail re-directs a mass of air of the apparent wind backwards to create a backward force (Force = ma). The reactive equal and opposite forward force pushes the sailboat ahead.

According to Galileo a boat can sail downwind faster than the wind for the same reasons that it can sail upwind faster than the wind. Each tack is simply the mirror image of the other one. This is because in both situations the sailboat experiences a headwind, and the true wind is moving backwards relative to the boat.

When sailing both upwind and downwind, the sail extracts momentum and energy from the true wind by slowing it down. That's it.

CONTENTS

| I | INTRODUCTION | 9 |
| II | NEWTONIAN ARGUMENT SUMMARIZED | 15 |

PART 1 BACKGROUND

| III | BACKGROUND – SAILING BASICS | 27 |
| IV | BACKGROUND – AIRFLOW ANALYSIS | 31 |

PART 2 NEWTONIAN MECHANICS

V	NEWTON APPLIED TO SAILING	39
VI	EXAMPLE CALCULATION	48
VII	SAILING INTO WIND	52
VIII	SAILING CONUNDRUM SOLVED	60
IX	ADDITIONAL ANALYSIS	63

PART 3 GALILEAN RELATIVITY

X	ALL MOTION IS RELATIVE	71
XI	DOWNWIND FASTER THAN THE WIND	74
XII	BLACKBIRD LAND YACHT ENIGMA	85

PART 4 FALSE THEORIES OF SAILING

XIII	FALSE THEORIES OF SAILING	91
XIV	FLUID MECHANICS	93
XV	FALSE VECTOR BASED SOLUTIONS	98
XVI	COMPARISON TO KITESURFING	106
XVII	A SAIL IS NOT A WING	110

PART 5 CONCLUDING REMARKS

XVIII	DISCUSSION	121
XIX	CONCLUSIONS	124
XX	ADDITIONAL INFORMATION	125
XXI	REFERENCES	126

I
INTRODUCTION

A. THE BASICS

THE NEWTONIAN APPROACH is significant as it provides new and useful insights. It challenges the prevailing sailing solutions based on fluid mechanics and resolves long-running nautical conundrums, including how a boat can sail downwind faster than the wind.

According to Newtonian mechanics, the sail re-directs a mass of air each second (m/dt) from the apparent wind towards the boat's stern at a velocity (dv) relative to the boat. The re-directed air decelerates when it interacts with the undisturbed apparent wind to create turbulence to create a backwards force (Force $_{BACK}$ = ma = m/dt x dv). The reaction generates an equal and opposite force that pushes the boat ahead. See Fig. 1a.

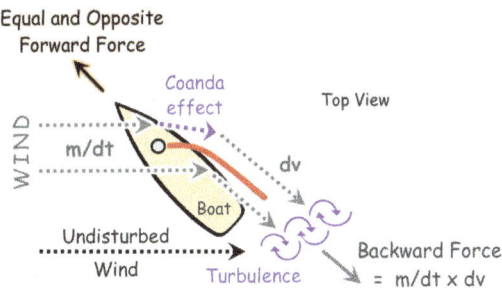

Fig. 1a. Airflows and Newtonian forces

A boat can sail upwind faster than the wind because:

- The apparent wind is re-directed by the sail against the wind itself to generate a force, which is a moving mass and not a fixed or static point. In contrast, existing theories rely on the apparent wind pushing directly against the sail to create a force.

- The apparent wind speed (dv) is only one factor that determines the force generated by the sail, and therefore, the speed of the sailboat. For example, if the sail size doubled, then the sail could double the mass of air re-directed each second (2 x m/dt). In turn, this would double the forward force generated (2 x Force = 2m/dt x dv).

- The Coanda effect on the leeward side of the sail increases when sailing closer into the wind. This effect increases the mass of air re-directed each second (m/dt), and therefore, the forward force generated (Force = m/dt x dv).

- A positive feedback loop arises: the boat's increased speed sailing into wind, causes an increase in 'm/dt' and 'dv' of the apparent wind, helped by the Coanda effect. This increases the forward force (Force = m/dt x dv), and therefore, the boat's speed increases further. Turning closer into the wind is like pressing down on the accelerator pedal of a car.

Then applying the ideas of Galileo, a boat can sail downwind faster than the wind for the same reasons that it can sail upwind faster than the wind. Each tack is simply the mirror image of the other one. This is because in both situations the sailboat experiences a headwind, and the true wind is moving backwards relative to the boat. See Fig. 1b.

When sailing downwind and upwind, momentum is transferred from the apparent wind to the boat via the sail by slowing the apparent wind down, which creates turbulence.

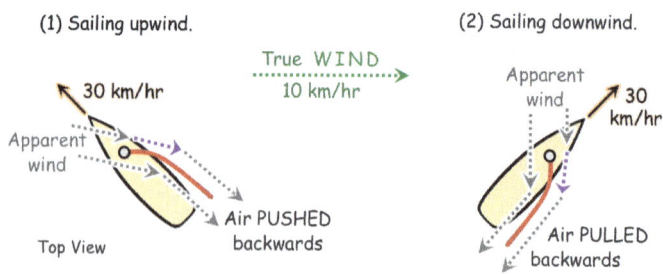

Fig. 1b. Apparent wind sailing

B. THE PHYSICS OF SAILING INTO THE WIND IS UNRESOLVED

NO EXPERIMENT ON a sailboat in realistic conditions has proven any theory or equation to be true. Problematically, many people mistake mathematical proof (e.g. Navier-Stokes equations) and computer simulations (e.g. CFD) for scientific proof. The failure to adequately explain sailing is puzzling given how long people have been sailing and how important sail-powered boats were to early global economic development.

A number of different theories have been proposed. The current preferred theories fail to explain sailing because they are based on a few **erroneous and unproven assumptions** that include: See Fig. 1c.

- Fluid mechanics (hydrodynamics) and Navier-Stokes equations can explain the forces created by a sail. [1][2][3][4]
- A sail creates a force by the wind pushing directly against it, as compared to the sail re-directing the airflow to push against the undisturbed wind.
- A sail has a similar design, shape, and functions to an airplane wing. Therefore, a sail produces similar airflows and forces as compared to a wing. In particular, false theories propose that the sail generates lift perpendicular to the sail. One obvious problem with this argument is that experts have not proven how a wing generates lift. [72]
- Vector-based solutions attempt to explain how the lift generated perpendicular to the sail is channeled into a forward force due to the interaction of the hull, keel, and sail, which allows a boat to sail into the wind.

However, analysis refutes these assumptions by showing that fluid mechanics fails to explain the forces on a sail; a sail is not a wing; and the main force created is not perpendicular to the sail. Past research failed to explain sailing downwind faster than the wind because they relied on these false assumptions.

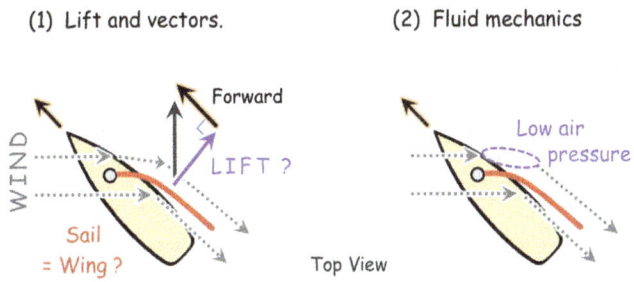

Fig. 1c. False vector-based solutions and fluid mechanics

C. NEWTONIAN MECHANICS

APPLYING NEWTONIAN MECHANICS to explain sailing is not a new concept. However applying Newtonian mechanics to sailing based on the wind being redirected and being explained using the mass flow rate is a novel approach.

It should not be surprising that Newton's Laws of Motion can explain the physics of sailing and that Galilean relativity helps explain sailing downwind faster than the wind. This is accepted physics being applied to explain how a sailboat moves.

The Newtonian approach provides significant new and useful insights, which solves several nautical conundrums, including:

- A boat's speed increases significantly when it is turned into the wind, despite having less sail area exposed to the wind.
- A boat sails more efficiently with multiple sails, instead of a single large sail with the same total sail area.
- A boat sails into the wind faster than the wind, but it sails directly with the wind at the speed of the wind.
- A boat can sail downwind faster than the wind. In turn, this analysis also solves the enigma of the how the Blackbird land yacht travels downwind faster than the wind.

Additional benefits of the Newtonian approach include:

- A simple and easily understood explanation of sailing that is consistent with accepted physics.
- A method for accurately calculating the forces, kinetic energy, and power generated by the sail.
- An explanation of apparent wind sailing.
- A correct understanding of the physics of sailing allows for improved methods of sail design and racing techniques.

Existing theories of sailing fail to provide such explanations and benefits. Therefore, considerable resources are wasted pursuing false theories.

If the physics of how a boat floats can be explained as simply as Archimedes did over 2,000 years ago, then the physics of sailing into the wind should be equally as straightforward to explain.

D. A HISTORICAL PERSPECTIVE OF SAILING

EUROPEANS WERE USING boats to trade for well over 3,000 years before the Portuguese perfected sailing into the wind in the 15th century, with sailboats such as the Caravel. [6] There is evidence that Polynesians managed this feat using catamarans (outriggers) in the Pacific thousands of years before the Portuguese. [7] See Fig. 1d.

This means that ancient Phoenicians, Greeks, Romans, Egyptians and Vikings relied on oar-power, water currents, and sailing with the wind to power their boats. In short, they rowed to cross the seas. At the same time, the Polynesians were effortlessly sailing across enormous distances of the Pacific Ocean using their knowledge of sailing into the wind. See Fig. 1e.

Fig. 1d. Polynesian Hokulea and Portuguese Caravel [29]

Fig. 1e. Ancient Trireme and Viking longboat [31]

Sailing into the wind was a pivotal technological development that allowed for accelerated global economic growth seen since the 15th century. Sea travel became significantly cheaper and faster, allowing for an expansion of global trade. This technology allowed the Portuguese to develop an empire based on trading with more distant parts of the world, including the spice trade with India. [6] See Fig. 1f.

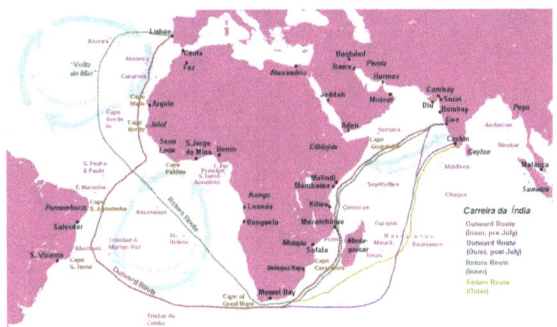

Fig. 1f. Portuguese global trade routes in the 1500s [30]

The rest of Europe quickly adopted the new sailing techniques, expanding their trade and colonial empires. Without this technology, the global economy and politics would have evolved very differently.

II

NEWTONIAN ARGUMENT SUMMARIZED

A. NEWTONIAN MECHANICS EXPLAINS SAILING

THE GENERATED FORWARD force depends primarily on the amount and speed of air re-directed by the sail. Boats sailing into the wind on a close haul at a positive sail angle-of-attack (AOA) re-direct a moving mass of air each second (m/dt) from the apparent wind backwards at a velocity relative to the boat (dv), which is helped by the Coanda effect on the leeward side.

As the re-directed air pushes against the undisturbed air from the apparent wind causing the wind to decelerate. This action results in turbulence and creates a backwards force (Force = ma = m/dt x dv). The reaction generates an equal and opposite forward force that pushes the boat ahead. See Fig. 2a-i and 2a-ii.

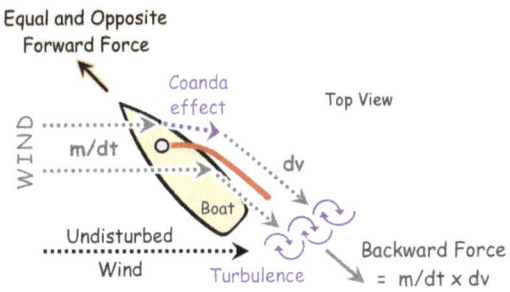

Fig. 2a-i. Newtonian forces acting on a sailboat

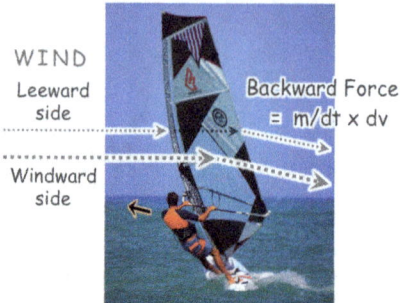

Fig. 2a-ii. Newtonian forces acting on a windsurfer [32]

B. MULTIPLE SAILS

THE NEWTONIAN APPROACH can explain why multiple sails (e.g. jib and mainsheet) provide a greater forward force than a single, large sail with the same total sail area. Multiple sails increase the mass flow rate (m/dt) without significantly jeopardizing the relative acceleration of air (dv). A higher 'm/dt' increases the force generated (Force = m/dt x dv). See Fig. 2b-i and 2b-ii.

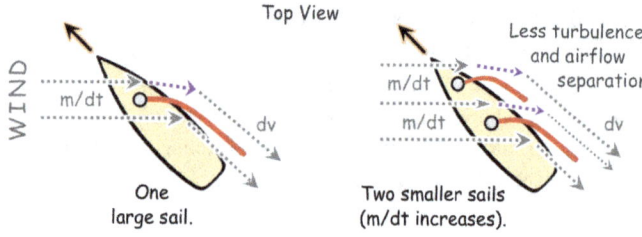

Fig. 2b-i. One large sail v. two smaller sails with the same total sail area

Fig. 2b-ii. Ship with mutiple sails [33]

C. EXAMPLE CALCULATION

THE EXAMPLE CALCULATION below demonstrates the Newtonian approach applied to sailing. See Fig. 2c-i.

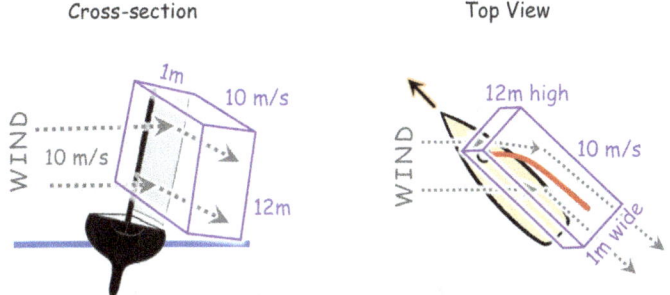

Fig. 2c-i. Volume of air re-directed by the sail

It is assumed that:

- Apparent wind is 10 m/s.
- Air density is 1.2 kg/m^3. [1]
- The sail is 12 m high.
- Sail reach of 1 m; which means the wind is re-directed 0.5 m on either side of the sail, facing the direction of the apparent wind.

The sail displaces a volume of air each second of 120 m^3/s:

Volume/dt = 12m x 1m x 10 m/s = 120 m^3/s

The displaced volume of air equals a mass of 144 kg/s of air:

m/dt = Volume/dt x Air Density
= 120 m^3/s x 1.2 kg/m^3
= 144 kg/s

If the apparent wind is re-directed by the sail at 8 m/s (dv), then according to Newtonian mechanics this creates a backward force of 1,152 N. The reaction generates an equal and opposite forward force of 1,152 N. See Fig. 2c-ii.

Force $_{BACK}$ = m/dt x dv
= 144 kg/s x 8 m/s
= 1,152 N
= Force $_{FORWARD}$

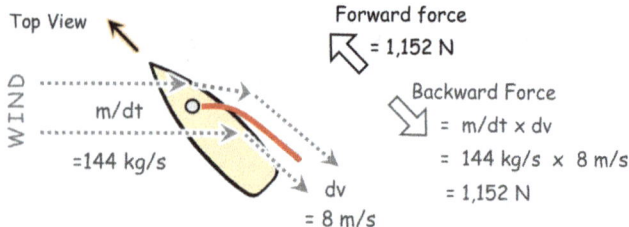

Fig. 2c-ii. Example calculation of Newtonian forces

D. SAILING INTO THE WIND

ACCORDING TO NEWTONIAN mechanics the closer the boat sails into the wind, the greater the 'm/dt' and 'dv'. Therefore, the greater the forces generated (Force = m/dt x dv), and the higher the boat's speed. This feat occurs because the force generated by the sail is not limited to the speed of the wind. In addition, when sailing closer into wind a positive feedback loop develops, accelerating the sailboat.

The Newtonian explanation in more detail is due to:

- The force generated by the sail pushes against the turbulence behind the sail, and not directly against the sail as commonly believed. See Fig. 2d-i.

- The force (Force = m/dt x dv) generated by the sail depends on the mass of air re-directed each second (m/dt) from the wind and its relative velocity (dv).

- When sailing into the wind, a significantly greater airflow (m/dt) is re-directed due to the Coanda effect on the leeward side of the sail at a higher

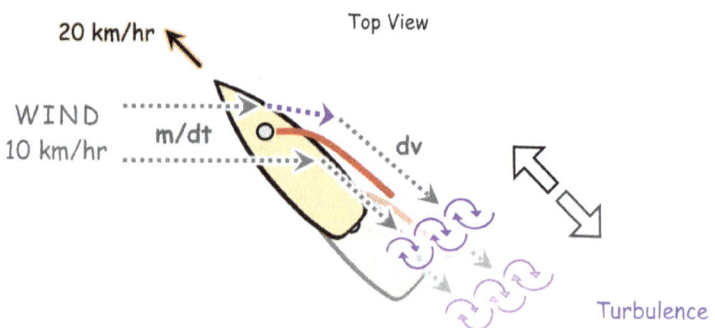

Fig. 2d-i. Sailing into the wind

Downwind faster than the wind

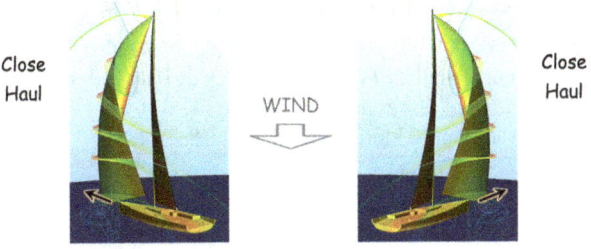

Fig. 2d-ii. Computer simulations of the Coanda effect [28]

velocity (dv). This generates a greater force (Force = m/dt x dv). See Fig. 2d-ii and 2d-iii.

- A positive feedback loop arises as the boat sails closer into the wind. As the boat's speed increases sailing into the wind, this causes an increases the 'm/dt' and 'dv' of the apparent wind, helped by the Coanda effect. Therefore the forward force generated (Force = m/dt x dv). In turn, this causes the boat's speed to increase further.

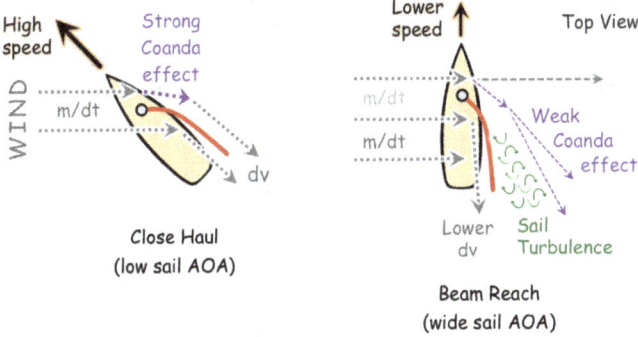

Fig. 2d-iii. Coanda effect is stronger on a close haul

E. CONUNDRUM: SAILING INTO THE WIND V. WITH THE WIND

A NAUTICAL CONUNDRUM exists: A sailboat on a close haul can sail into the wind faster than the wind itself but a boat sailing directly with the wind cannot sail faster than the wind. For example, a boat can sail at 20 km/hr into a 10 km/hr wind, but not faster than 10 km/hr with the wind. See Fig. 2e.

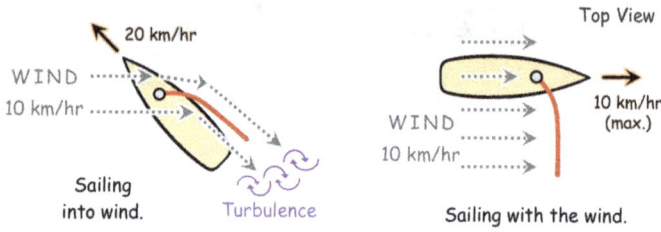

Fig. 2e. Example of sailing speeds relative to the wind

This is a conundrum for several reasons:

- At first glance it appears illogical for a boat to be able to sail into the wind and to do so faster than the wind. Surely, the wind should blow the boat in the direction of the wind. How then does the sailboat use the wind to create a forward force in the opposite direction to the wind, which is greater than the wind itself?
- Compared to running with the wind, on a close haul relatively little sail is exposed to the wind. Surely, the sail would generate less power?

Newtonian mechanics and the Coanda effect provide a simple solution to this conundrum:

- As explained above, a boat can sail upwind faster than the wind because (i) the force generated by the sail pushes against the turbulence behind the sail, and (ii) the force generated by the sail depends on the mass of air re-directed each second (m/dt) from the wind and its relative velocity (dv) compared to the boat.
- When running with the wind, a sail has nothing to push against and cannot re-direct the wind. Therefore, no 'dv' and no equal and opposite forces are possible. The boat's speed is then limited to the speed of the wind. In addition, no Coanda effect is possible on the leeward part of the sail, limiting the amount of wind being re-directed. Only the windward side of the sail is providing any forward force. See Fig. 2e.

F. APPARENT WIND SAILING DOWNWIND FASTER THAN THE WIND

NEW TECHNOLOGIES HAVE produced high-performance sailboats that can sail on almost any tack at speeds many multiples (e.g. 3 - 6 times) faster than the true wind, both downwind and upwind. This is called apparent wind sailing. See Fig. 2f-i.

Fig. 2f-i. Apparent wind sailing [46]

Apparent wind sailing downwind faster than the wind has eluded explanation until now. It is argued in this book that the same physics that explains sailing into the wind faster than the wind also explains sailing downwind faster than the wind.

It is counter intuitive for a boat to sail downwind faster than the wind, when powered by the wind. The conundrum is that in apparent wind sailing boats sail downwind several times faster than the wind. For example, a boat can sail downwind on a broad reach at a boat speed of 30 km/hr, with a 10 km/hr tailwind. See Fig. 2f-ii.

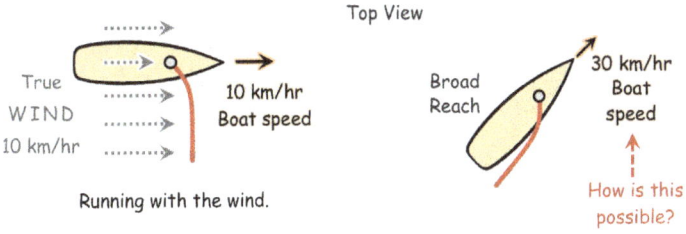

Fig. 2f-ii. Sailing downwind faster than the wind

G. SOLUTION TO SAILING DOWNWIND FASTER THAN THE WIND

KEY ISSUES FOR apparent wind sailing downwind faster than the wind include:

- **A sailboat maintains the apparent wind in front of it**, regardless of whether it is sailing upwind or downwind.

- **A boat sails downwind faster than the wind,** which is explained by combining Newtonian mechanics with Galileo's assertion that all motion is relative.

When sailing upwind and downwind, the sails re-direct the relative airflow (headwind) backwards toward the stern of the boat to create a force (Force = m/dt x dv) that propels the boat ahead.

When sailing downwind and upwind faster than the wind, a boat always experiences a headwind. At this point, the true wind is moving backwards relative to the boat in both cases. According to Galileo it is indistinguishable to a sailor whether the wind is pushing or pulling against the sails. Without external reference points a sailor doesn't know whether they are sailing downwind or upwind. A boat can sail downwind faster than the wind for the same reasons that it can sail upwind faster than the wind. Sailing downwind faster than the wind is simply the mirror image of sailing upwind faster than the wind. Nonetheless, to avoid confusion in this analysis, for a boat sailing upwind the wind is described as being pushed backwards against the boat. Conversely, for a boat sailing downwind faster than the wind, the wind is described as being pulled backwards relative to the boat.

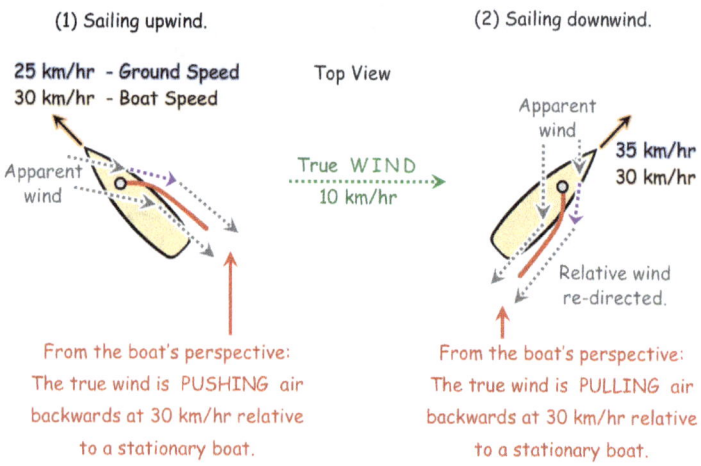

Fig. 2g-i. Relative movement of the wind and boat - 1

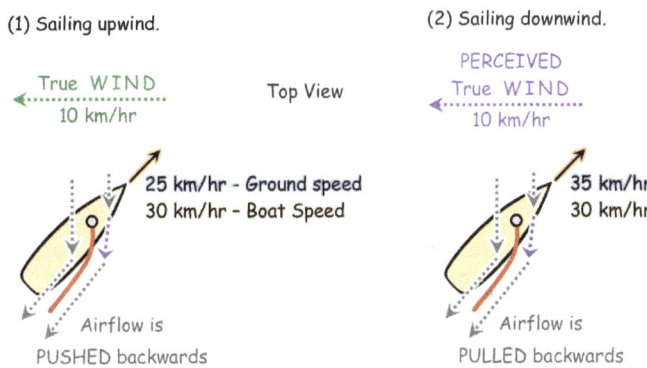

Fig. 2g-ii. Relative movement of the wind and boat - 2

For example, with a true wind of 10 km/hr, if two boats have the same boat speed of three times the true wind (i.e. 30 km/hr) traveling upwind and downwind; then these two boats have different ground speeds. This example is illustrated in a different way with the same boats side-by-side and true wind blowing in opposite directions. The relative action (pushing / pulling) of the wind is also opposite. See Fig. 2g-ii.

From the perspective of the boat sailing downwind, the true wind is perceived as travelling in the opposite direction (i.e. backwards).

This aspect of the argument is confusing at first, as it is counter intuitive. Yet it breaks no laws of physics and fits with what is observed in practice. It is hard to reason that a tailwind is pulling air backwards relative to the boat sailing downwind. The boat is outrunning its tailwind and accelerating further away. It is simply the reverse of sailing into the wind faster than the wind.

The different relative ground speeds and directions observed on the two different boats in this example are largely irrelevant to the forces generated by the sail.

- **The transfer of momentum from wind to the sail**

According to Newtonian mechanics, a force is created when the sail redirects the relative airflow backwards against the undisturbed true wind, creating turbulence. The turbulence provides something for the re-directed wind to push against and generates the equal and opposite forward force. This process slows down the true wind, extracting momentum from it, which is transferred to the boat.

Therefore, the sail transfers momentum from the wind to the boat by slowing down the true wind, both during upwind and downwind sailing. The wind is the only source of power and energy for the sailboat. The transfer of momentum helps explain how a boat can sail downwind faster than the wind.

- **Transition to apparent wind sailing**

    The transition from running with the wind to apparent wind sailing downwind using example speeds is described as follows: See Fig. 2g-iii.

    - Drag from the water generally means running with the wind producing a speed below that of the wind.

    - Shifting to a broad reach alters the AOA of the apparent wind, which exposes the leeward side of the sail to the apparent wind. This change boosts the mass of air re-directed backwards each second (m/dt), which is helped by the Coanda effect. In turn, this increases the generated force (Force = m/dt x dv) by the sail, causing the boat to accelerate.

- As the boat gains speed, the AOA of the apparent wind moves towards the front of the boat. The boat transitions to apparent wind sailing and the physics allow the boat to exceed the speed of the wind.

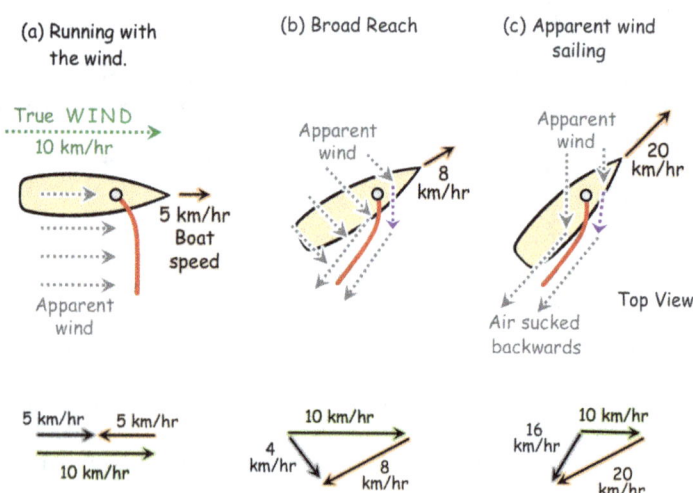

Fig. 2g-iii. Downwind sailing with example speeds

PART I

BACKGROUND

III

BACKGROUND SAILING BASICS

A. THE BASICS OF SAILING

ONLY THE BASICS of sailing relevant to this book are outlined below. The sail AOA is the angle between the sail's direction and the apparent wind. This is different to the boat's AOA, which is the angle between the boat's heading and the apparent wind. See Fig. 3a-1.

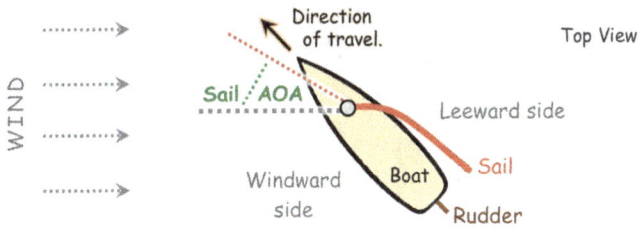

Fig. 3a-i. Sailing basics

The primary functions of the key controls include:

- The sail provides the power for the boat.
- The rudder controls the direction of the boat.
- The keel provides stability by preventing the boat from being tipped over and pushed downwind. The weight of the keel also helps to keep the boat's centre of mass lower and close to the water level.

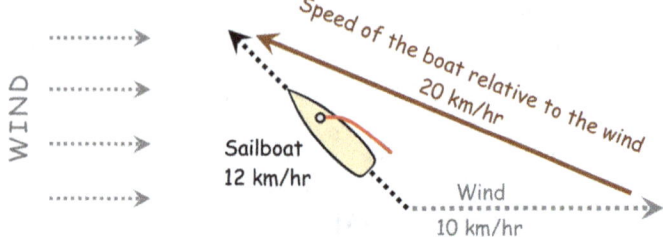

Fig. 3a-ii. Illustration of relative speeds

Sailboats can sail into and with the wind faster than the wind itself. For example, a sailboat sailing at 12 km/hr ground speed on close haul into a 10 km/hr wind, could be travelling at 20 km/hr relative to the wind. Similarly, the same feat is possible sailing with the wind. See Fig. 3a-ii.

The speed and direction of any water current relative to the boat can cause drift by pushing against the boat's hull and keel. However, this is largely irrelevant to the forces generated by the sail. See Fig. 3a-iii.

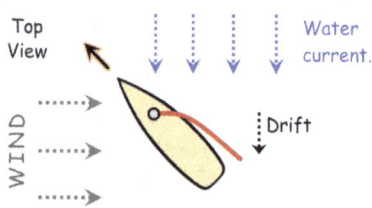

Fig. 3a-iii. Drift due to water current

B. POINTS OF SAIL AND BOAT SPEED

TO MOVE FORWARD, the force generated by the sail must be sufficient to overcome the drag from the water and any drift due to the water current. Drag is the force required to push the water out of the boat's path due to the inertia from the water.

The boat's direction relative to the wind and sail alignment is shown in Fig. 3b. The points of sail show a boat's speeds are roughly symmetrical with respect to the wind.

- The boat achieves the same speeds whether the boat is heading to the east or west, when facing a northerly wind. The boat can travel faster than the wind when sailing both into and with the wind.

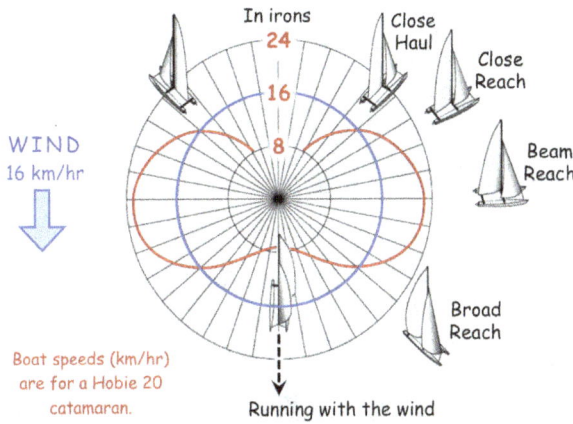

Fig. 3b. Points of sail and example boat speeds [69]

- The exception to the symmetry is that a boat cannot sail directly into the wind. But when running with the wind, directly downwind, the boat is limited to true wind speed.
- The symmetry of the boat speeds with respect to the wind indicates the involved forces are universal and constant. The same forces affect how the boat sails upwind and downwind.

For a boat sailing into the wind, apparent wind speed is the wind measured relative to the boat. Apparent wind speed can be quite different from true wind speed, as experienced by a stationary observer.

C. CATAMARANS DO NOT NEED A KEEL

CATAMARANS SAILING INTO the wind without a keel demonstrate the keel is not always required to oppose the sideways force. Instead, two separate hulls are used to prevent the boat from tipping over, and to provide resistance to the sideways force, for two reasons:

- The leeward hull in the water provides resistance to the sideways force, preventing the catamaran from being pushed downwind, like a boat with a keel.
- The catamaran's second hull on the windward side provides balance and prevents the boat from tipping over. Like an outrigger it uses gravity to replicate the benefits of a keel. See Fig. 3c-i and 3c-ii.

Fig. 3c-i. Catamarans sailing into the wind [35]

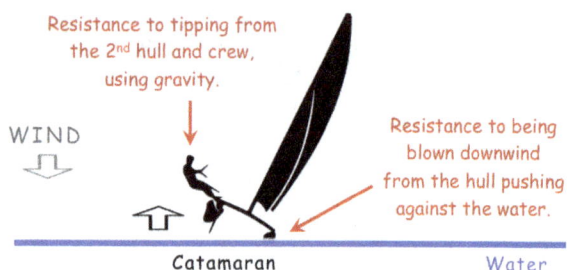

Fig. 3c-ii. The forces on a catamaran sailing into the wind

D. THE FORCES CREATED BY A RUDDER

THE SAME NEWTONIAN principles apply to how a rudder is used to steer a boat. For a sailboat that is moving forward, changing the AOA of the rudder re-directs water flow under the boat. This action creates a force based on the mass of water re-directed each second and the relative velocity of the water flow. The reaction generates an equal and opposite force, which pushes the stern (rear) of the boat in the direction caused by these forces. The stern of the boat is simply pivoted around the boat's centre of gravity. See Fig. 3d.

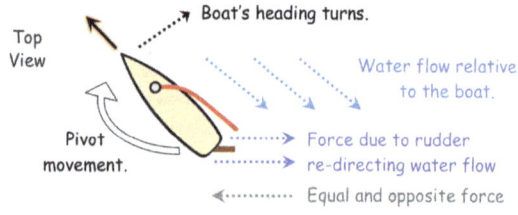

Fig. 3d. Rudder creates a force by re-directing water flow

IV

BACKGROUND AIRFLOW ANALYSIS

A. TWO SAIL AIRFLOWS

THERE ARE TWO separate key airflows on the sail: See Fig. 4a-i

- On the windward side, the sail acts as a physical barrier and re-directs (pushes) the windward airflow backwards, which creates high air pressure. This is similar to how any barrier re-directs the airflow or the path of a rubber ball. See Fig. 4a-ii.
- The forward motion of the sail causes a vacuum of low air pressure to arise on the leeward side of sail. This is similar to how low pressure arises behind any moving object. The area of low-pressure pulls the leeward airflow back-

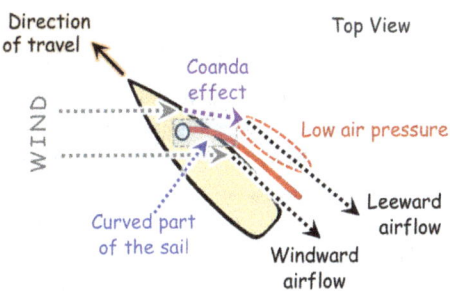

Fig. 4a-i. Wind re-directed by the sail

wards along the curved (bowed/concaved) surface of the leeward side of the sail, helped by the Coanda effect. The sail is effectively pulling the air around the sail, bending its direction. See Fig. 4a-iii.

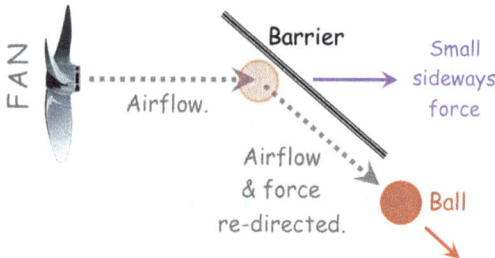

Fig. 4a-ii. Airflow re-directed by a barrier (windward side)

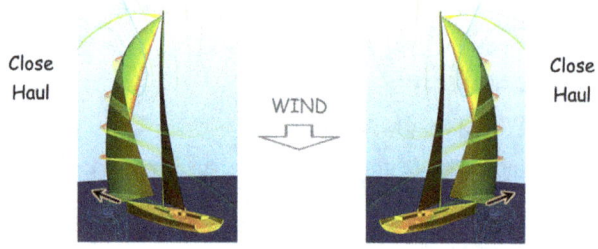

Fig. 4a-iii. Computer simulations [28] of the Coanda effect (leeward side)

B. THE COANDA EFFECT DEMONSTRATED BY A SPOON

FLUID FLOW NATURALLY follows a curved surface due to the Coanda effect. For example, water falling from a tap is re-directed by the curved shape of a spoon, as can be demonstrated by a simple classroom experiment. See Fig. 4b.

According to Newtonian mechanics, altering the direction of the fluid flow creates a force (Force = ma = m/dt x dv). The reactive equal and opposite force pushes the spoon diagonally upwards. The Coanda effect can have a significant effect on fluid flow (m/dt), and therefore, the generated force.

In contrast, if the spoon's orientation is reversed so the curvature faces the

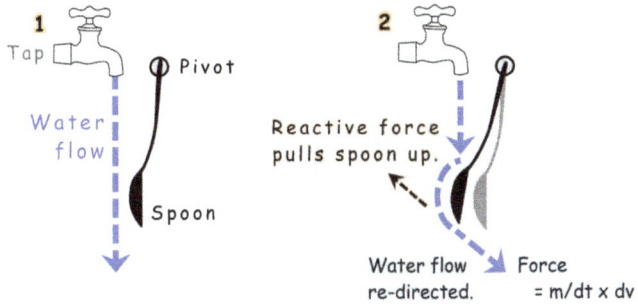

Fig. 4b. Coanda effect – Spoon experiment

opposite direction, or if other non-curved objects, such as a knife, are placed in the water flow while connected to a pivot. Then these objects do not generate a Coanda effect and are immediately pushed back to their original position by the water flow from the tap. They do not remain suspended in the water flow.

These observations are only achieved if the water flow is sufficiently fast and voluminous, and the spoon is sufficiently large with enough curvature. Otherwise, the spoon does not remain attached to the water flow.

Similar experiments with curved objects, such as balls, have produced comparable results. However, researchers made different explanations and conclusions regarding the forces involved. [16]

C. THE COANDA EFFECT ON AIRPLANE WINGS

DUE TO THE lack of data and wind tunnel experiments available for sails, airplane wings are used as a proxy to demonstrate airflows and the Coanda effect. See Fig. 4c-i.

Some fighter jet wing and fuselage designs show pronounced curvature that maximizes the Coanda effect. See Fig. 4c-ii.

The amount of air re-directed by the Coanda effect depends on the maintenance of laminar (smooth / non-turbulent) airflow on the topside of the wings to maximize the amount of air displaced down each second (m/dt). See Fig. 4c-iii.

In general, wings produce a stronger Coanda effect with laminar airflow at a lower AOA, higher airspeed, and where the wings are deepest (largest chord, such as near the fuselage). Conversely, the Coanda effect is weakest at high AOA, slower air-

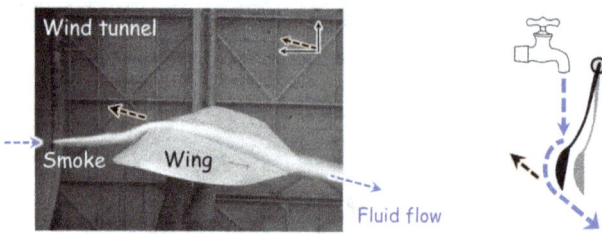

Fig. 4c-i. Coanda effect on an airplane wing and spoon [36]

speeds, and where the wings are narrow (small chord, such as at the wing tips). The flat undersides of wings are typically designed to push air down without inducing any Coanda effect.

According to Newtonian mechanics, the optimum wing AOA maximizes the combined airflow re-directed by both the underside and topsides of the wing, and therefore, the force generated.

The top airflow is the sensitive to changes in wing AOA due to the Coanda effect. Whereas the lower airflow does not rely on the Coanda effect making it more stable and less sensitive to changes in the wing AOA. This dynamic is evidenced by stalls arising due to disrupted airflow on the topside of wings. This means that when analyzing airflow that causes lift, attention is focused on the upper airflow.

Fig. 4c-ii. Curved wing and fuselage designs of jets [37][38]

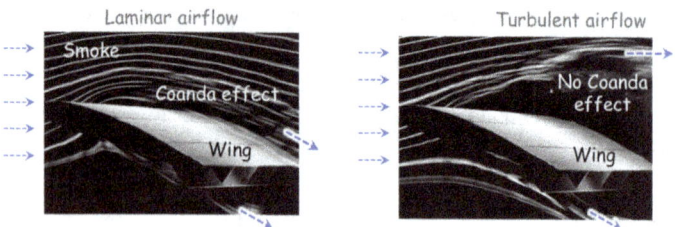

Fig. 4c-iii. Smooth vs. turbulent airflows on a wing [39]

D. TELLTALES

TELLTALES ARE SMALL pieces of string attached to both sides of a sail. See Fig. 4d. Sailors use telltales as airflow indicators to judge what type of airflow is being maintained on the different parts of the sail. This helps sailors assess how well the sail is aligned to the optimum sail AOA under the prevailing conditions. In short, telltales provide evidence that the sail is re-directing the air as desired.

The equivalent to a telltale on an airplane wing is a stall indicator, which measures when airflow across the wing is being disrupted sufficiently to generate a stall.

For example, disrupted telltales on the leeward sail indicate turbulent airflow. Reducing the sail AOA could establish laminar airflow and reduce turbulence. This is like an airplane wing that has a high AOA and turbulent airflow on the topside of the wing. A typical solution to reduce the turbulent airflow is to reduce the wing AOA by lowering the nose of the airplane.

Fig. 4d. Telltales on a sail [40][41]

E. THE COANDA EFFECT AND SAIL SHAPE

THE OPTIMUM SAIL size and shape depends on the boat's size, dimensions, weight, etc. as well as the purpose of the sailboat (e.g. racing, leisure etc.). A balance between speed, stability and maneuverability is sought.

A bowed or concave sail shape enhances the Coanda effect on the leeward side of the sail but reduces the effectiveness of the windward side of the sail. A rigid (flat) sail has the reverse effect, which reduces the re-directed wind on the leeward side of the sail, while enhancing the re-directed wind on the windward side of the sail. Therefore, the optimum sail shape is a balance between these two parameters. A lot of equipment is employed while sailing to prevent the sail from losing its shape, and thus, to maintain the ideal bowed sail shape. This equipment includes batons, reefing points, backstays, etc. See Fig. 4e.

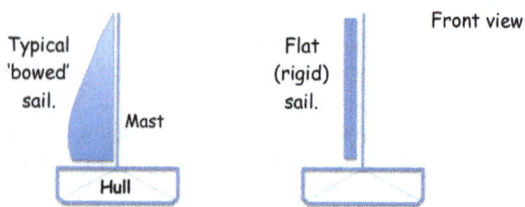

Fig. 4e. Different sail shapes

This book postulates that the leeward side of the sail can re-direct more wind (displace more air) than the windward side when sailing into the wind for two reasons:

(i) The typical bowed shape of a sail enhances the Coanda effect and the sail reach on the leeward side.

(ii) The windward side's high-pressure system based on pushing air is less efficient at re-directing wind (displacing air) backwards, as compared to the low-pressure system of the Coanda effect on the leeward side of the sail that pulls air backwards. The Coanda effect has a greater reach, and thus, affects air a greater distance away from the sail.

PART 2

NEWTONIAN MECHANICS

V

NEWTON APPLIED TO SAILING

A. SUMMARY

THE PHYSICS OF sailing into the wind is straightforward. The sail re-directs the apparent wind backwards towards the stern of the boat, resulting in two forces: See Fig. 5a-i.

- A large backward force created by re-directing the wind on both sides of the sail against the atmosphere behind the boat. The reaction is the forward force, which causes the boat to move forwards through the water.
- A small sideways force exerted by the windward airflow pushing directly against the sail, causing the boat to tilt.

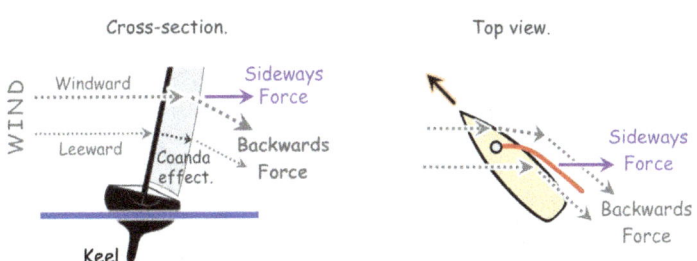

Fig. 5a–i. Two forces acting on a sail on a close haul

On a close haul into the wind, the sail re-directs most of the wind's momentum and energy backwards. Only a relatively small but significant force is applied sideways on the sail due to the windward airflow, which is resisted by the boat's hull and keel. See Fig. 5a-ii.

- The keel and hull limits drift and prevents the boat from being blown downwind.
- The keel limits tilting and prevents the sailboat from tipping over.

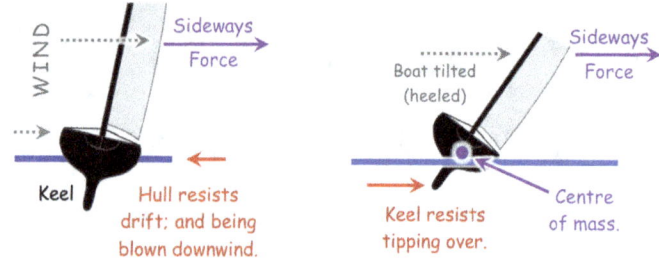

Fig. 5a-ii. Resistance from the hull and keel

B. FORWARD FORCE = M/DT X DV

THE TWO AIRFLOWS re-directed by the sail collide with the undisturbed apparent wind behind the sail, causing them to decelerate and creating turbulence. The turbulence provides something for the re-directed airflow to push against to generate a forward force. According to Newtonian mechanics, the mass of wind re-directed by the sail each second (m/dt) and its relative velocity (dv)

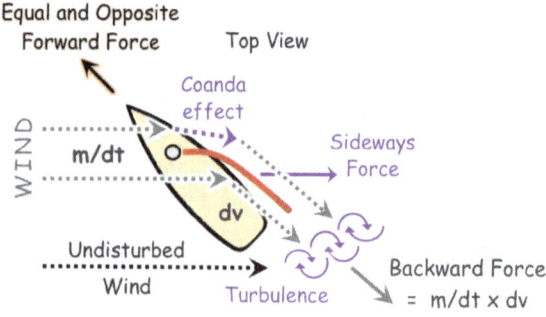

Fig. 5b-i. Newtonian forces acting on a sailboat

creates a backward force (Force $_{BACK}$ = ma = m/dt x dv). The reaction generates an equal and opposite forward force, pushing the boat ahead (Force $_{FORWARD}$ = Force $_{BACK}$). In short, the sail re-directs air backwards, which pushes the boat forward. See Fig. 5b-i and 5b-ii.

Fig. 5b-ii. Newtonian forces acting on a windsurfer [32]

Where:

    m = Mass of air re-directed by the sail.
    m/dt = Mass flow rate.
    dt = Change in time (per second).
    dv = Change in velocity (v) of re-directed air.
    v = Velocity of the wind re-directed by the sail.
    a = dv/dt = Acceleration.
    Force = ma = m x dv/dt = m/dt x dv [1]
    Force = ma = m x dv/dt = d(m/v)/dt [1]
    Momentum = mv [1]

The physics described above are summarized by the following equations:

| Force $_{BACK}$ | = | ma | = | m/dt x dv | (1) |
| Force $_{BACK}$ | = | Force $_{FORWARD}$ | | | (2) |

Equations (1) and (2) can be combined as follows:

Force $_{BACK}$	=	Force $_{FORWARD}$	=	m/dt x dv	(3)
Or simply:		Force $_{FORWARD}$	=	m/dt x dv	(4)
Units: N			=	kg/s x m/s	

In equation (4), the change in velocity of the air is expressed as 'dv,' and not as acceleration ('dv/dt'). This action is due to a one-off force (impulse) from the sail, which is not time dependent. Whereas the mass of air flown through by the sail is time dependent, and is expressed as 'm/dt'.

C. THE MOMENTUM THEORY OF SAILING

THERE IS NO net gain or loss of momentum, energy, or mass in the process of sailing. The apparent wind has significant momentum and kinetic energy due to its mass and velocity. Momentum and energy are transferred from the wind to the boat via the sail, by the sail slowing down the wind in two ways: See Fig. 5c.

- On direct contact with the sail, the wind slows down, as it is re-directed by the sail (from V_1 to V_2 in Fig. 5c). This action creates a sideways force on the sail, which can be measured by the deceleration of the wind and the mass of the wind that is deflected by the sail. The sideways force does not contribute directly towards the forward force that pushes the boat ahead. The momentum transferred to the sail by the sideways force is relatively limited if the boat sails efficiently.

- The re-directed apparent wind exits the sail and interacts with the undisturbed apparent wind, to create turbulence and the backward force. Again, the momentum transferred can be calculated based on the mass of the wind re-directed and its deceleration (from V_2 to V_3 in Fig. 5c).

The transfer of momentum is expressed by the following equation:

$$\text{Force}_{BACK} = ma = d(mv)/dt \quad (7)$$

The forward and backwards forces are equal, as described by equation (2), and therefore, equation (7) can be re-stated as follows:

$$\text{Force}_{FORWARD} = ma = d(mv)/dt \quad (8)$$

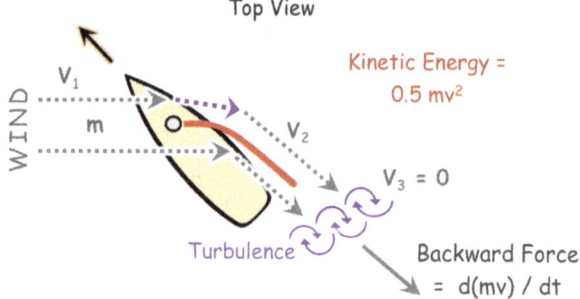

Fig. 5c. Newtonian forces due to change in momentum

D. THE TWO NEWTONIAN EQUATIONS

THE ANALYSIS PROVIDES two Newtonian equations for the forward force generated by a sail:

$$\text{Force}_{FORWARD} = ma = m/dt \times dv \text{ (mass flow rate)} \quad (4)$$
$$\text{Force}_{FORWARD} = ma = d(mv)/dt \text{ (momentum theory)} \quad (8)$$

Both equations (4) and (8) are based on Newtons 2nd Law of Motion (Force = ma). Both equations are correct and produce the same values but express the same factors differently.

E. ADDITIONAL COMMENTS

THE TRANSFER OF kinetic energy from the wind to the sail can be expressed by the equation:

$$\text{Kinetic Energy} = 0.5 \, mv^2 \, [1] \quad (6)$$

This is significant because the parameters to calculate the forward force generated while sailing are consistent with those used to calculate the momentum and kinetic energy involved. This is not the case for the forces calculated based on fluid mechanics. See Fig. 5c.

Additional considerations in applying Newtonian mechanics to sailing include:

- The sail size, AOA, shape, height, width, and apparent wind speed affect 'm/dt' and 'dv', and therefore, the force created.
- The difference between velocity of the apparent wind and 'dv' represents the force, energy and momentum lost to pushing the boat sideways, friction or otherwise.
- 'dv' is a weighted average of the 'dv' for the two airflows on the windward and leeward sides of the sail.
- 'm/dt' and 'dv' vary with the distance away from the sail and are different on the leeward and windward parts of the sail.
- This approach permits analysis of how the force generated (Force = m/dt x dv) changes by making adjustments that alter 'm/dt' and 'dv'. For example, analysis of how changes to the sail's AOA alter 'm/dt' and 'dv'.

F. AIRBOAT COMPARISON

A SAIL ACTS in a manner that is similar to a big propeller that pushes air backwards on an airboat. See Fig. 5f.

The comparison between a sailboat and an airboat is only meant to illustrate the principles of Newtonian mechanics involved. The airboat pushes air backwards, which generates an equal and opposite forward force. A sail re-directs the apparent wind to create a force like that of an airboat. To be clear, this is the limit of the comparison being made.

Fig. 5f. Forces acting on an airboat [42]

G. SAIL REACH

SAIL REACH IS the horizontal distance away from the sail where the sail affects the air. It is a key parameter that determines the amount of wind that the sail can re-direct, and therefore, the mass flow rate (m/dt).

Sail reach depends on the sail size, height, width, shape, and AOA, as well as the apparent wind speed and the Coanda effect on the leeward side of the sail. Sail reach may not be equidistant on either side of the mast, and therefore, it may be different on the windward and leeward sides. See Fig. 5g.

Downwind faster than the wind

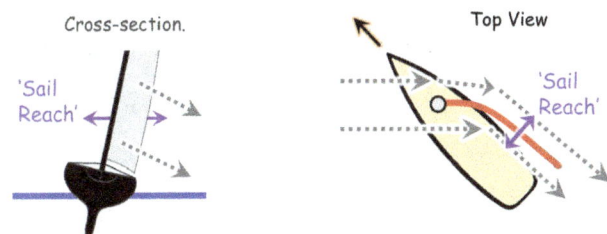

Fig. 5g. Sail reach

H. SAIL AREA, HEIGHT, AND WIDTH

THE TERMINOLOGY USED for sail area, height, and width is illustrated in Fig. 5h.

There is no simple relationship between the sail area exposed to the wind and the forward force generated by the sail. However, an analysis of sail height and width provides a method to estimate the generated forward force by looking at how these factors affect 'm/dt' and 'dv'.

On a close haul into the wind:

- Sail height (or mast height) primarily determines the amount of air caught by the sail (m/dt). A taller sail increases the 'm/dt', as a larger area of sail faces the apparent wind.

  This is a similar principle to how a glider generates lift with its long wingspan. The glider flies through a large mass of air each second (m/dt), which it re-directs to create lift when soaring.

- Sail width primarily determines the relative acceleration of the air caught (dv). A wider sail maximize the 'dv' achieved.

  This explains why a significant forward force can be generated on a close haul into the wind despite relatively little of the sail area being exposed to apparent wind.

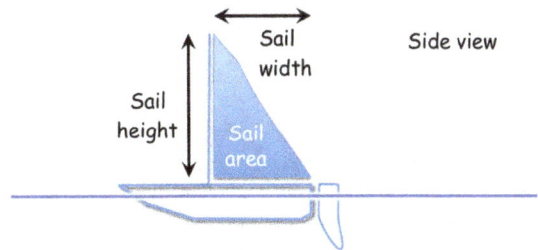

Fig. 5h. Sail area, height, and width

I. SAILBOAT TILT

THE TILT OF a sailboat affects the sail airflows, but it does not significantly affect the amount of apparent wind that the sail passes through each second (m/dt). For example, a 12 m tall mast has 12 m exposed to the wind regardless of the boat's tilt (pitch). See Fig. 5i.

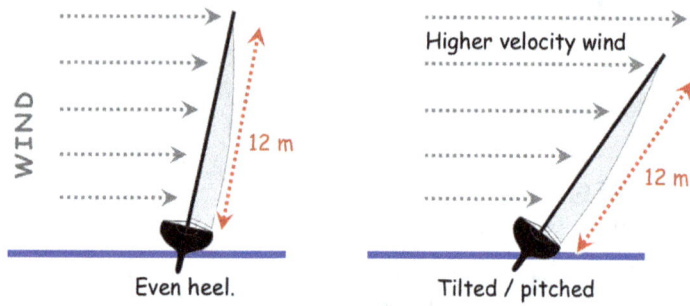

Fig. 5i. Sailboat on an even heel and tilted

However, tilt can also be important to the airflows and forces generated where:

- The wind velocity varies a lot with altitude (height above the water).
- Wind spills over top of the sail, and therefore, less apparent wind is re-directed backwards each second (m/dt).

J. NEWTONIAN MECHANICS IS UNIVERSAL

NEWTONIAN MECHANICS BASED on the mass flow rate provides a universal theory of motion that can be applied universally to all forms of motion achieved by pushing a fluid backwards. In other words, the physics of how a sail generates a force is similar to how other objects create a forward force by pushing air backwards against the atmosphere.

Compare a sail, a jet engine and a propeller. A sail re-directs airflow (apparent wind) against the atmosphere to generate a force. Air is mostly gaseous nitrogen and oxygen. A jet engine pushes exhaust gases, which are mostly carbon dioxide and water vapor, downward against the atmosphere to generate upward thrust. A propeller accelerates a mass of air backwards against the atmosphere to generate forward thrust. See Fig 5j-i.

Fig. 5j-i Newtonian forces acting on a propeller and jet engine [71]

In all cases for a sail, jet engine and propeller; thrust or the forward force is calculated using the Newtonian mechanics based on the mass flow rate (Thrust = m/dt x dv). [1][73][75]

By way of additional illustration, the thrust (Thrust = m/dt x dv) generated by a jet pack can also be calculated based on the mass of water each second (m/dt) accelerated downward to a velocity (dv) by the jet pack relative to the person. See Fig 5j-ii.

Fig. 5j-ii. Newtonian forces acting on a jet pack [68]

VI
EXAMPLE CALCULATION

A. OVERVIEW

AN EXAMPLE CALCULATION of the forward force generated by a sail from a boat sailing into the wind based on Newtonian mechanics and the mass flow rate is provided below.

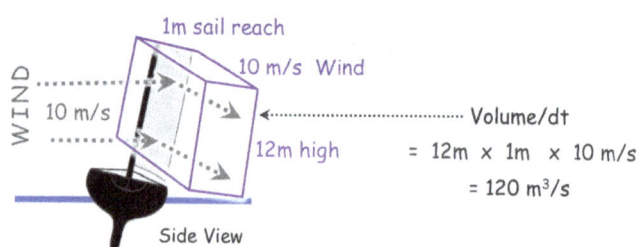

Fig. 6a. Sail dimensions

B. ASSUMPTIONS

THE KEY ASSUMPTIONS include: See Fig. 6a.

- Standard air density is 1.2 kg/m^3. [1]
- Apparent wind speed is 10 m/s (36 km/hr) or V^1.
- 'dv' is 8 m/s (approximately 29 km/hr), or V_2.
- The sail is 12 metres high.
- Sail reach is 0.5 metres horizontally on either side of the sail, (1.0 metres total).

For simplicity, several situational assumptions are made in this example:

- The sailboat is sailing at a constant velocity into a laminar, constant velocity wind, with no water currents. Therefore, the boat is not accelerating.
- The sail airflows that generate a force are assumed to be rectangular in shape.
- Vortices or wind shear are not significant.
- The sailboat is assumed to be sailing almost vertically, with little heel (tilt / pitch).
- The sail width is not considered to be directly relevant to this calculation. It is indirectly important, as it affects the sail reach and the 'dv' that is assumed.

There is a difference between the velocity of the apparent wind of 10 m/s (V_1) and the speed at which the re-directed air leaves the sail (dv) at 8 m/s (V_2). This represents the force, energy, and momentum lost to pushing the boat sideways, friction or otherwise.

C. EXAMPLE CALCULATION – METHODOLOGY

IN THIS EXAMPLE, the volume of air displaced each second by the sail is estimated. Standard air density translates this volume into a mass of air displaced each second (m/dt). Then considering the assumed value for 'dv', the backwards force can be calculated.

- **The volume of air displaced by the sail is estimated to be 120 m^3/s** based on the sail height, sail reach, and apparent wind speed.

Volume of air displaced per second	=	Sail height	x	Sail Reach	x	Apparent wind speed
120 m^3/s	=	12 m	x	1 m		x 10 m/s

A 12 metre vertically high sail with a 1.0 m sail reach (which re-directs the wind 0.5 m on either side of the sail), sailing into a strong 10 m/s (36 km/hr) apparent wind displaces a volume of air of 120 m³/s. See Fig. 6a.

- **The mass of air displaced by the sail each second (m/dt)** is calculated to be 144 kg/s based on the volume of air displaced and the standard air density of 1.2 kg/m³.

    m/dt   =   Volume/dt x Air Density
    144 kg/s =   120 m³/s x 1.2 kg/m³

- **The average velocity of the air displaced (dv) by the sail is assumed to be 8 m/s.** This reflects how efficiently the sail re-directs apparent wind of 10 m/s. This is the velocity the airflow leaves the sail to interact with the undisturbed apparent wind, creating turbulence.

- **The backward force (Force $_{BACK}$)** is calculated from multiplying 'm/dt' by 'dv', using the standard Newtonian equation:

    Force $_{BACK}$ =   m/dt x dv
                    =   144 kg/s x 8 m/s
                    =   1,152 kg m/s²
                    =   1,152 N
                    =   Force $_{FORWARD}$

The backward force of 1,152 N generates an equal and opposite forward force (Force $_{FORWARD}$), as shown above.

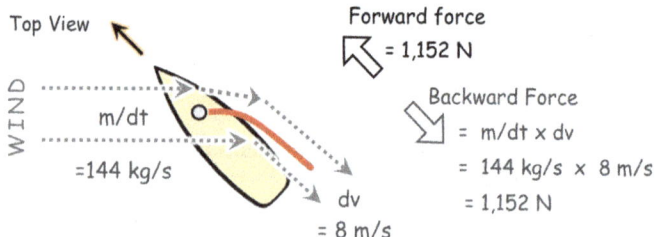

Fig. 6c. Example calculation of Newtonian forces

According to Newtonian mechanics, an airflow of 144 kg/s of air, at a velocity relative to the boat of 8 m/s, creates a backwards force of 1,152 N. The reaction generates a forward force of 1,152 N that pushes the boat ahead. See Fig. 6c.

D. COMMENTS

THIS EXAMPLE IS an approximation and for illustration purposes only, which is intended to demonstrate how Newtonian mechanics can be applied in practice. It is not meant to be precisely accurate nor overly realistic. Nonetheless a more detailed and realistic calculation would eventually need to include more accurate assumptions.

The straightforwardness of the Newtonian approach is illustrated by the simplicity of this example calculation.

The key parameters that are the most difficult to estimate in practice are the 'sail reach' and 'dv'. These assumptions are the most speculative of all the assumptions in this example calculation.

Other theories of sailing, such as fluid mechanics and false vector-based approaches, fail to provide simple example calculations based on realistic conditions.

VII

SAILING INTO WIND

A. SAILING INTO THE WIND AND AOA

BOATS ARE OBSERVED to sail upwind faster than the wind. As the boat sails closer into the wind and its sail AOA decreases, the generated forward force increases. This causes the boat's speed to also increase in a non-linear manner. Boats sail significantly faster as the AOA declines. This occurs up to a point when the sails stall while sailing directly into the wind, and cease to generate a forward force. See Fig. 7a-i.

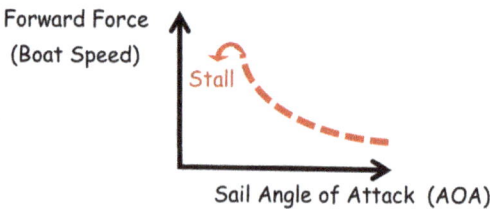

Fig. 7a-i. Sail AOA and forward force

High-performance boats are observed to sail up to several multiples of the wind's speed (3 - 6 times faster). This occurs despite having considerably less sail area directly exposed to the wind on the windward side of the sail.

When sailing directly into the wind (in irons), the wind pushes the sail relatively flat. The sail is not re-directing any wind (no m/dt and dv arise), and therefore, no forward force is generated (Force = m/dt x dv). See Fig. 7a-ii.

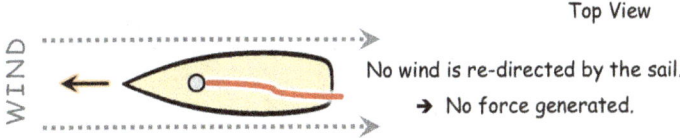

Fig. 7a-ii. Sailing directly into the wind

B. THE NEWTONIAN EXPLANATION

ACCORDING TO NEWTONIAN mechanics the closer the boat sails into the wind, the greater the 'm/dt' and 'dv'. Therefore, the greater the forces generated (Force = m/dt x dv), and the higher the boat's speed. This feat occurs because the force generated by the sail is not limited to the speed of the wind. In addition, when sailing closer into wind a positive feedback loop develops, accelerating the sailboat.

The Newtonian explanation in more detail is due to:

- The force generated by the sail pushes against the turbulence behind the sail, and not directly against the sail as commonly believed. See Fig. 7b.

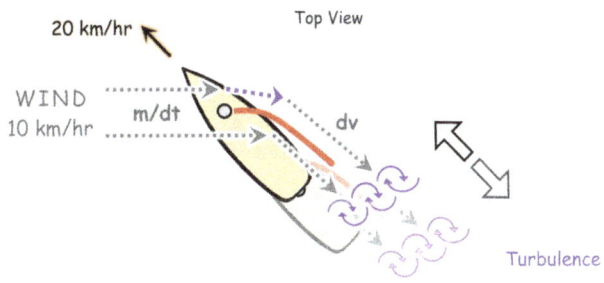

Fig. 7b. Sailing into the wind

- The force (Force = m/dt x dv) generated by the sail depends on the mass of air re-directed each second (m/dt) from the wind and its relative velocity (dv) compared to the boat.

  In other words, the apparent wind speed is only one factor that determines the force generated by the sail, and therefore, the speed of the sailboat. For example, if the sail height and size doubled, then the sail could double the mass of air re-directed (2 x m). In turn this would double the forward force generated (2x Force = 2ma).

- When sailing into the wind, a significantly greater airflow (m/dt) is re-directed due to the Coanda effect on the leeward side of the sail at a higher velocity (dv). This generates a greater force (Force = m/dt x dv). This aspect is explained below in more detail.

The increase in airflow on the leeward side of the sail more than offsets any reduced effectiveness of the windward side of the sail that has less sail exposed to the wind.

C. COANDA EFFECT AND AOA

ACCORDING TO NEWTONIAN mechanics, the optimum sail position is one that maximizes the combined airflow re-directed by both the windward and leeward sides of the sail.

The airflow on the leeward side of the sail is the most sensitive to changes in

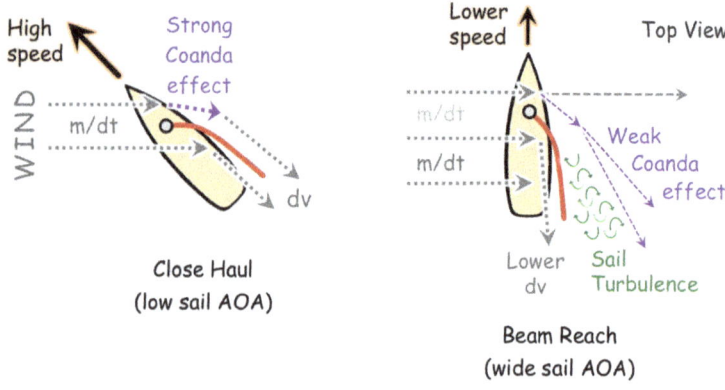

Fig. 7c-i. Coanda effect is stronger on a close haul

sail AOA, as compared to the windward side. Sailing closer into the wind with a lower sail AOA increases 'm/dt' and 'dv', and therefore, the forward force generated as well as the boat's speed. This occurs because: See Fig. 7c-i.

- On a beam reach, more sail on the windward side is directly exposed to the wind, increasing 'm/dt' on the windward side. However, the sail AOA is much greater, so the apparent wind is re-directed through a much greater angle, reducing 'dv' significantly. More of the apparent wind's force is lost to pushing the boat sideways, rather than forward.

- The Coanda effect on the leeward side of the sail is highly sensitive to the sail's AOA. On a close haul, there is less turbulence and airflow separation, and therefore, more air displaced (higher m/dt and dv) on the leeward part of the sail.

   When close hauled, the lower sail AOA deflects the apparent wind at a lower angle. Therefore, it loses less of its force and maximizes 'dv'.

- A positive feedback loop arises as the boat sails closer into the wind. The mass

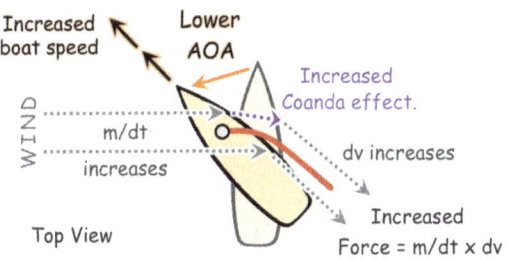

Fig. 7c-ii. Lower sail AOA feedback loop

of air sailed through each second (m/dt) increases due to the Coanda effect on the leeward side of the sail. This causes an increase in the generated forward force (Force = m/dt x dv), and therefore, the boat's speed. See Fig. 7c-ii.

As the boat's speed increases, 'm/dt' and 'dv' increase further, causing a greater increase in the forward force generated (Force = m/dt x dv). In turn, this pushes the boat ahead at a higher speed. See Fig. 7c-iii.

The impact of the positive feedback loop on the boat's speed is greatest at low speeds. This is because drag on the hull from the water is proportional to the boat's velocity squared. Therefore, as the boat's speed increases to a high velocity, drag increases rapidly.

The exponential nature of drag explains why hydrofoils are essential for any high-speed sailboat. Hydrofoils minimize drag by raising the hull above the water.

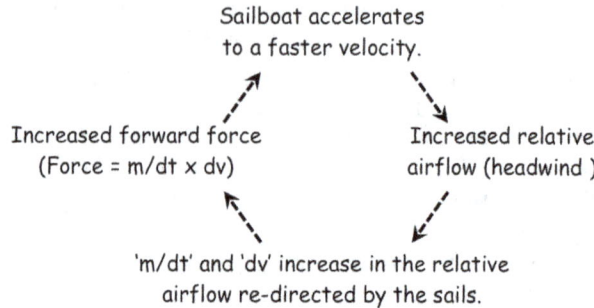

Fig. 7c-iii Positive feedback loop sailing closer into the wind

- The sailboat heels (tilts) less and moves more efficiently through the water with less drag than when on a close haul.
- The Coanda effect is greatest when sailing closest to the wind, but not directly into or with the true wind. See Fig. 7c-iv.

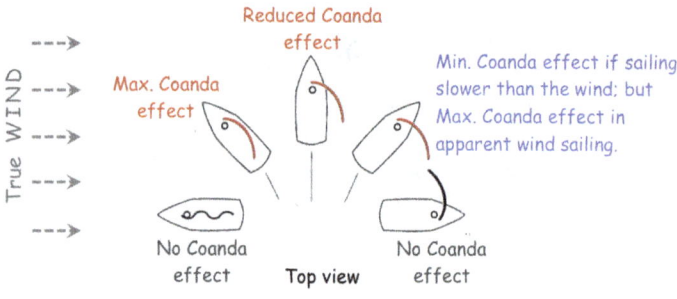

Fig. 7c-iv. The Coanda effect and sail AOA

In addition, when sailing downwind, the Coanda effect on the leeward side of the sail increases significantly once the boat starts sailing faster than the wind. This change is due to the apparent wind shifting towards the bow (front) of the boat as the boat's speed increases. See Fig. 7c-ii.

- The leeward airflow experiences little turbulence, as it is protected by the sail from the apparent wind. This aspect allows the leeward side to displace

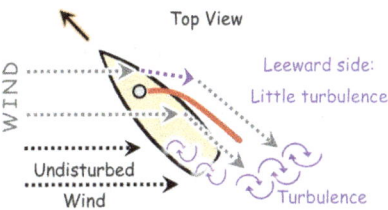

Fig. 7c-v. Little turbulence on the leeward side

the air backwards more effectively, as compared to the windward side. See Fig. 7c-v.

In summary, turning a sail into the wind is like pressing down on the accelerator pedal of a car. The sail is more efficient at creating a forward force as it re-directs a greater mass of air each second (m/dt) on the leeward side at a higher relative velocity (dv). This is despite reduced 'm/dt' on the windward side on a closer haul.

D. DRAG AND MOMENTUM TRANSFERRED TO THE WATER

The drag experienced by the boat's hull (Drag $_{HULL}$) is described by applying the standard equation for drag: [1]

Drag $_{HULL}$ = 0.5 (Velocity^2 x Water Density x Surface Area
 x Drag Coefficient)

Where:

Velocity = Boat speed through the water.

Surface Area = Surface area of the boat's hull facing the direction of travel.

Drag Coefficient = The drag coefficient of the boat's hull in the water.

This equation means that if the boat's velocity doubles, then the drag from the water against the boat's hull quadruples, assuming all other variables are constant.

The explanation for this relationship is provided by Newtonian mechanics based on the mass flow rate (Force = ma = m/dt x dv). The force exerted by the boat's hull (Force $_{HULL}$) can be calculated using the mass of water travelled through each second (m $_{WATER}$ /dt) and the velocity (dv $_{WATER}$) that this water is accelerated away from the hull. The reactive equal and opposite force is called drag. In other words, drag is the resistance from the water due to its inertia, against boat's hull physically pushing the water out of its path. Therefore, the physics of drag can be expressed by the Newtonian equation: See Fig. 7d.

$$\text{Force}_{HULL} = m_{WATER}/dt \times dv_{WATER}$$
$$= \text{Drag}_{HULL}$$

If the boat's velocity doubles, it travels through twice the mass of water each second than before (2x m $_{WATER}$ /dt). In addition, the boat's momentum has also doubled. Hence the hull accelerates the water passed away from it at twice the velocity as before (2x dv $_{WATER}$). Combined, these two aspects mean that the force exerted by the boat's hull against the water, and therefore drag, quadruples: See Fig. 7d.

$$4 \times \text{Force}_{HULL} = (2 \times m_{WATER}/dt) \times (2 \times dv_{WATER})$$
$$= 4 \times \text{Drag}_{HULL}$$

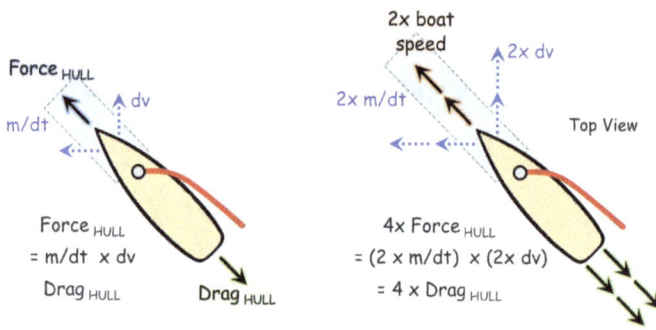

Fig. 7d. The force exerted by the hull on the water and drag

In addition, the boat moves at a constant velocity when the forward force generated by the sails from the apparent wind (Forward Force $_{SAIL}$), equals the drag experienced by the boat's hull as it moves through the water due to resistance from the water. In this situation of stable forward motion, the sails re-direct a mass of the air each second (m $_{AIR}$ /dt) at a velocity (dv $_{AIR}$) backwards, which equals the mass and velocity of the water displaced sideways by the boats hull each second. Momentum is conserved in the process. This dynamic can be summarized by the following equation:

Forward Force $_{SAIL}$ = Drag $_{HULL}$

$m_{AIR} / dt \times dv_{AIR}$ = $m_{WATER} / dt \times dv_{WATER}$

A key point is that Newtonian mechanics based on the mass flow rate can explain the exponential nature of drag due to increases in the boat's velocity, as well as the conservation of momentum in the sailing process. Whereas fluid mechanics cannot explain either. In addition, the use of Newtonian mechanics to explain the standard equation for drag is new. It has not been presented elsewhere previously.

VIII
A SAILING CONUNDRUM SOLVED

A. SAILING INTO / WITH THE WIND

WHEN SAILING WITH the wind, the wind pushes the sail and the boat downwind. However, when sailing into the wind, the sail re-directs apparent wind backwards to generate a forward force. See Fig. 8a.

Fig. 8a. Boats sailing into and with the wind [43]

B. CONUNDRUM DESCRIBED

A SAILBOAT ON a close haul can sail into the wind faster than the wind itself. However, a boat sailing with the wind cannot sail faster than the wind. Many people claim that a boat can sail several multiples faster than the wind speed, which is true for sailboats with hydrofoils. On the other hand, a boat cannot sail directly into the wind (in irons). See Fig. 8b.

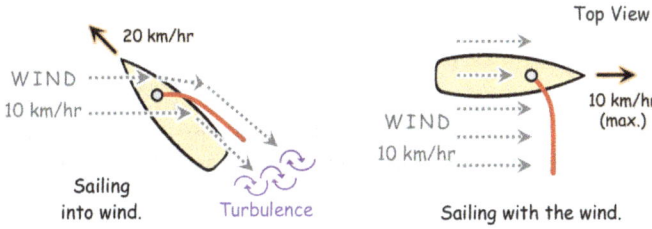

Fig. 8b. Example of sailing speeds relative to the wind

This is a conundrum for several reasons:

- At first glance it appears illogical for a boat to be able to sail into the wind and do so faster than the wind itself. Surely the wind should blow the boat in the direction of the wind. How then does a sailboat use the wind to create a force in the opposite direction that is greater than the wind?
- The amount of sail area that is directly exposed to the wind does not determine the force generated by the sail. For example, on a close haul, relatively little sail is exposed to the wind but a lot of force is generated, as compared to sailing with the wind.

C. THE CONUNDRUM EXPLAINED

NEWTONIAN MECHANICS AND the Coanda effect provide a simple solution to this conundrum.

As explained above, the closer the boat sails into the wind, the greater the 'm/dt' and 'dv'. Therefore, the greater the forces generated (Force = m/dt x dv), the higher the boat's speed. The force generated by the sail is not limited to the speed of the wind. In addition, when sailing closer into wind a positive feedback loop develops, accelerating the sailboat

In contrast, when running with the wind, the sail has nothing to push against and cannot re-direct any wind. No 'dv' nor equal and opposite force are generated. Therefore, the boat's speed is limited to the speed of the wind. In addition, no Coanda effect is generated on the leeward part of the sail, limiting the amount of re-directed wind. Only the windward side of the sail is providing any forward force. See Fig. 8c-ii.

D. OTHER EXPLANATIONS

THE NEWTONIAN SOLUTION of sailing into the wind is consistent with other experts' views. For example: "For sailing downwind, one wants fairly square sails, which are best at catching the wind. But for sailing upwind, taller narrower sails are best, ..." [3]. According to Newtonian mechanics:

- Boats with tall narrow sails have a long mast area exposed to the wind in the direction of travel, which maximizes 'm/dt' due to the Coanda effect on the leeward side. This principle is similar to how the long narrow wings with high aspect ratios of albatrosses and gliders can generate lift by re-directing the wind when soaring. This wing design also maximizes the airflow displaced each second (m/dt) due to the Coanda effect.
- On the other hand, square sails maximize the wind caught by the windward side. In this case, this boat performs best sailing downwind, where there is no Coanda effect.

IX
ADDITIONAL ANALYSIS

A. MULTIPLE SAILS

THIS NEWTONIAN BASED analysis helps explain why multiple sails (e.g. jib and mainsail) provide a greater forward force than one large sail, despite the total sail size being the same.

Multiple sails increase the mass flow rate (m/dt) without significantly jeopardizing the relative acceleration of the air (dv). For example, a boat with two sails has two leading edges facing the wind, which increase the effective sail reach. This configuration also enhances airflow on the leeward side of the sail, reducing turbulence and airflow separation.

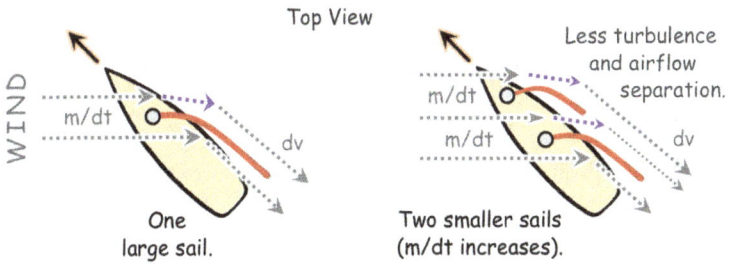

Fig. 9a-i. One large sail v. two smaller sails with the same total sail area

Fig. 9a-ii. Ship with multiple sails [34]

This effect increases the overall force generated using the Newtonian equation, Force = ma = m/dt x dv. In this case, the total sail size is the same for the boat with multiple sails as the boat with one sail. See Fig. 9a-i and 9a-ii.

A sail's 'dv' is limited to the speed of the apparent wind. Therefore, to maximize the generated forward force, a sailboat has no choice but to maximize 'm/dt'. One way of doing this is to increase the number of sails. Some additional considerations of the effect on the force generated by multiple masts include:

- The drawback to more sails is increased drag from the additional masts. There is a trade-off between increased 'm/dt' and increased drag from additional sails.

- The physical environment of a boat limits how many sails it can have. Larger boats tend to have more sails.

- In principle, additional sails on a boat are similar to the physics for additional wings on an airplane.

- Overlapping sails affect the airflows and consequently the forces involved. However, this aspect of sailing is beyond the scope of this book.

B. MOMENTUM, KINETIC ENERGY, AND POWER

MASS, MOMENTUM, AND energy are transferred from the wind to the sail, to power the boat forward. There is no net gain or loss of mass, momentum, and energy in this process.

If the apparent wind's velocity doubles, then its momentum also increases (momentum = mass x velocity). This aspect enables the wind to push the boat forward faster.

In addition, the analysis above is supported by the calculation of the kinetic energy produced by the sail to push the boat forward. The standard equation for kinetic energy is as follows:

Kinetic Energy (K.E.) = $0.5 \, mv^2$ [1]

Where:

m = Mass of air re-directed by the sail to generate a force.
v = Velocity of this mass of air re-directed by the sail.

The kinetic energy of the apparent wind is proportional to its velocity squared. Therefore, if the apparent wind's velocity doubles, then its kinetic energy quadruples. In turn, this explains why a boat's velocity increases in a non-linear manner with increased wind speed.

The power generated by the sail can then be calculated based on the amount of energy transferred from the wind to the boat, per unit time. These units are consistent with the units used to calculate the forces involved using Newtonian mechanics, including the mass and velocity of the wind. See Fig. 9b.

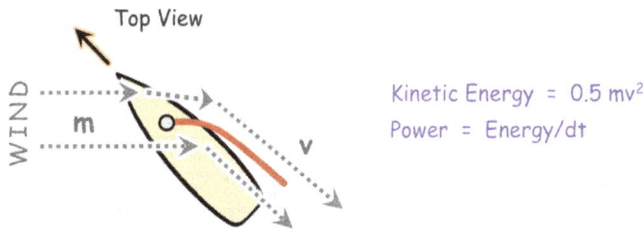

Fig. 9b. Force, kinetic energy, and power from a sail

No other theory of sailing provides such an accurate and simple method to calculate the forces, kinetic energy and power generated by a sail.

C. SAIL AOA AND BOAT ALIGNMENT

THE OPTIMUM SAIL AOA varies with the boat's heading, as shown in Fig. 9c. Sailors must adopt these configurations to sail as fast as possible. In this situation the forces generated are as close to the desired boat heading as possible.

More precisely, according to Newtonian mechanics, the ideal sail alignment (AOA) that maximizes 'm/dt' and 'dv', and therefore, the forces generated

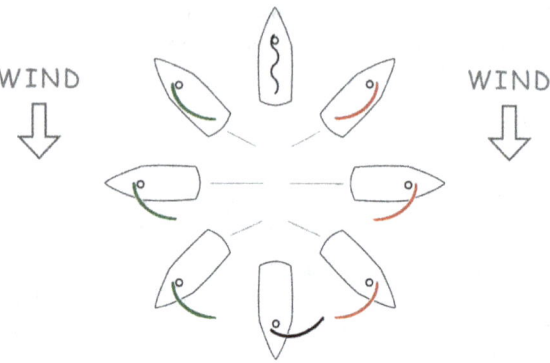

Fig. 9c. Points of sail [67]

(Force = m/dt x dv), changes with the boat heading. Maximizing 'm/dt' and 'dv' also means maximizing the combined parameters for 'm/dt' and 'dv' on both the leeward and windward sides of the sail.

D. SUB-OPTIMAL CONFIGURATIONS

IF THE SAIL AOA and boat's heading are not aligned, the force generated by the sail is not in the direction of travel, causing the boat to crab forward. However, this configuration is inefficient because it creates additional drag from the boat's hull moving through the water at an angle to the direction of travel. See Fig. 9d.

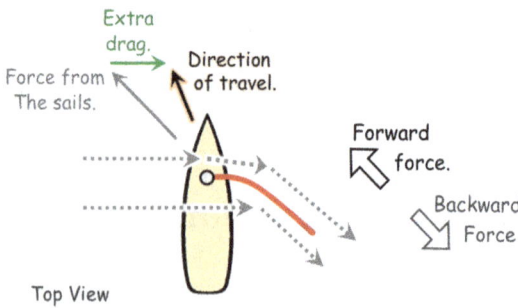

Fig. 9d. Sub-optimal configuration

Downwind faster than the wind

E. HEADING DIRECTLY INTO THE WIND

THE FOLLOWING EXAMPLE is of a sub-optimal configuration, where the boat adopts a close haul for the sail but points the bow into the wind. See Fig. 9e.

This configuration causes the boat to be pushed backwards, for the following reasons:

- The sail produces a forward force diagonal to the boat's heading.
- The hull and the keel provide little resistance to the oncoming wind, allowing the wind to push the boat backwards.
- The backwards motion by the boat reduces the 'm/dt' and 'dv' and therefore, the forward force generated (Force = m/dt x dv).
- The boat is sailing dead downwind but is facing upwind.

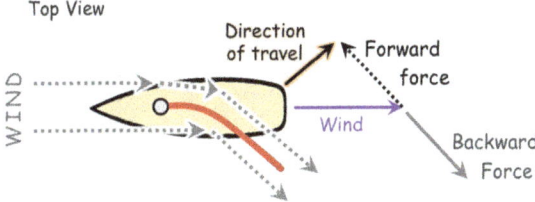

Fig. 9e. Sailing directly into the wind with a rigid sail

F. AN ALTERNATIVE PERSPECTIVE

TO PROVIDE AN alternative perspective, a diagram of forces acting on a sailboat is shown below with the wind from a different direction. The boat remains on a close haul tack, while the forces involved also remain the same. See Fig. 9f.

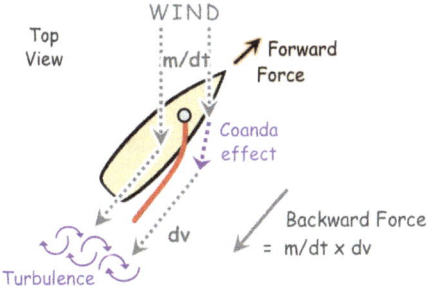

Fig. 9f. Forces acting on a sailboat – different perspective

PART 3

GALILEAN RELATIVITY

X
ALL MOTION IS RELATIVE

A. SKYDIVING EXAMPLE

ACCORDING TO GALILEAN relativity all movement is relative.

A skydiver cannot tell the difference between a free-fall decent at terminal velocity and indoor skydiving without external references. The relative airflow experienced by the skydiver is the same in both cases. It feels the same to the skydiver to fall through the air, as it does to have a big fan blow air upwards against the skydiver's body. The skydiver is floating in the air in both situations, even though these two processes may appear to be opposite to an observer. See Fig. 10a.

Free-fall skydiving.

Indoor skydiving.

Fig. 10a. Free-fall and indoor skydiving [47] [48]

B. SAILING DOWNWIND

A BOAT SAILING dead downwind, at the same speed as the wind, feels like it is dead in the water from the perspective of the sailboat, in the absence of buoys or external reference points. See Fig. 10b.

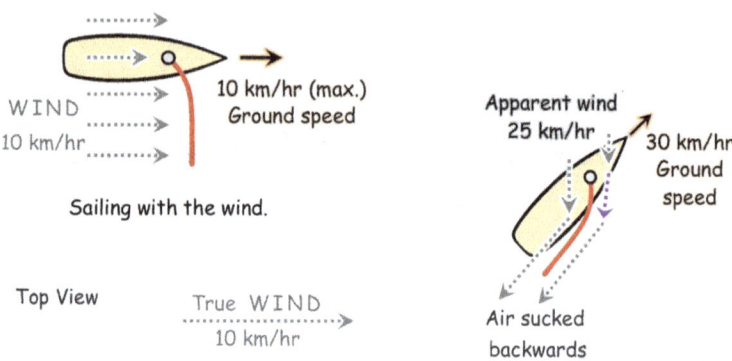

Fig. 10b. Sailing downwind

C. APPARENT WIND SAILING

APPARENT WIND SAILING is done at speeds many times faster than the true wind speed. The apparent wind is almost always in front of the boat. This means that whether the true wind is pushing (headwind) or pulling (tailwind) becomes irrelevant to how it generates a force.. When sailing upwind and downwind, the sail extracts momentum from the wind by slowing it down. Sailing upwind is almost a mirror image of sailing downwind, excluding external references (ground speed).

If a boat can extract momentum and energy from sailing into apparent wind, which is blowing (pushing) against its sails, then logically the boat can also extract momentum and energy from apparent wind sucking (pulling) against its sails. This applies as long as there is a wind for the sails to slow down.

This explanation is new and appears counter intuitive, making it conceptually confusing and difficult to accept. Therefore this point is elucidated further below.

In other words, in apparent wind sailing for boats sailing faster than the wind: the wind pushes against a boat sailing into the wind, but pulls against a boat sailing with the wind.

- Past analysis has overlooked that a tailwind is pulling against a boat sailing downwind.
- Sailing downwind is the mirror image of sailing upwind, both experience a headwind. The true wind is moving backwards from the perspective of the boat.
- There is a lack of alternative explanations for how a boat can sail downwind faster than the wind. It may appear counter intuitive, but it can be observed in practice.
- The sail behaves the same way when in a headwind, irrespective of whether it is sailing upwind or downwind. For example, the leeward side of the sail re-directs air and benefits from the Coanda effect, the same way as sailing into the wind.

The relative airflow is towards the stern (back) of the boat. The air is being pulled (or sucked) backwards relative to the boat and the sailboat behaves like it is in a headwind. The wind is not pushing against the sail, but it is being pulled backwards against the sail.

From the perspective of the boat sailing an angle downwind at 30 km/hr (headwind), with a 10 km/hr true wind (a tailwind), it can provide the boat with up to 40 km/hr ground speed. The boat experiences the wind moving backwards relative to itself, even though the wind is moving in the same direction. Because the wind is moving more slowly than the boat, it is described as pulling the airflow against the boat backwards.

From the perspective of the sailboat, Galilean relativity explains there is no distinction between a wind that is pulled or pushed in a given direction. For example, when sailing into an apparent wind, there is no difference between a wind that is pushed or pulled from the boat's perspective. The circular wind in a cyclone or anti-cyclone provides air moving from high to low-pressure zones in the atmosphere. It unclear whether the wind is pushed, pulled, or both.

XI

DOWNWIND FASTER THAN THE WIND

A. APPARENT WIND SAILING

NEW TECHNOLOGIES HAVE produced high-performance sailboats (e.g. catamarans, hydrofoils, iceboats and land sailing craft) with large and efficient sails combined with low-drag hulls (e.g. hydrofoils).

High-performance boats can sail on almost any course at speeds many multiples (3-6 times) faster than the true wind, both downwind and upwind. The new technologies have magnified the velocities achieved by a boat along the points of sail. The boat remains unable to sail into the wind and performance is poor when sailing directly downwind. See Fig. 11a.

Fig. 11a. High-performance sailing boats [44] [45]

Newtonian mechanics explains apparent wind sailing by how better technology has increased the 'm/dt' and 'dv' created by the sails, and therefore, the force generated (Force = m/dt x dv).

Better sails are more efficient at re-directing a greater mass of air backwards each second (m/dt) at a higher velocity (dv) relative to the boat. The force generated by the sails, and therefore, the speed of the boat, depends primarily on 'm/dt' and 'dv.' However, ultimately the generated force is limited by the speed of the true wind.

B. KEY ISSUES

THE KEY ISSUES of apparent wind sailing downwind faster than the wind include:

- **A sailboat maintains apparent wind in front of the boat**, regardless of whether it is sailing upwind or downwind.

- **A sailboat sails downwind faster than the wind**, which is explained by combining Newtonian mechanics with Galileo's assertion that all motion is relative.

  When sailing upwind and downwind, the sails re-direct the relative airflow (headwind) backwards toward the stern of the boat to create a force (Force = m/dt x dv) that propels the boat ahead.

  From the perspective of a sailor standing on the boat, the wind pushes against the boat sailing upwind and pulls (sucks) against the boat sailing downwind. However, the sailor and the boat cannot tell the difference between being pulled or pushed. These forces are indistinguishable.

  Nonetheless, to avoid confusion in this analysis, for a boat sailing upwind the wind is described as being pushed backwards against the boat. Conversely, for a boat sailing downwind faster than the wind, the wind is described as being pulled backwards relative to the boat.

  For example, with a true wind of 10 km/hr, if two boats have the same boat speed of three times the true wind (i.e. 30 km/hr) traveling upwind and downwind; then these two boats have different ground speeds. See Fig. 11b-i.

  The example can be shown in a different way with the same boats side-by-side and the true wind directions in opposite directions. The relative action (pushing/pulling) of the wind would also be opposite. See Fig. 11b-ii.

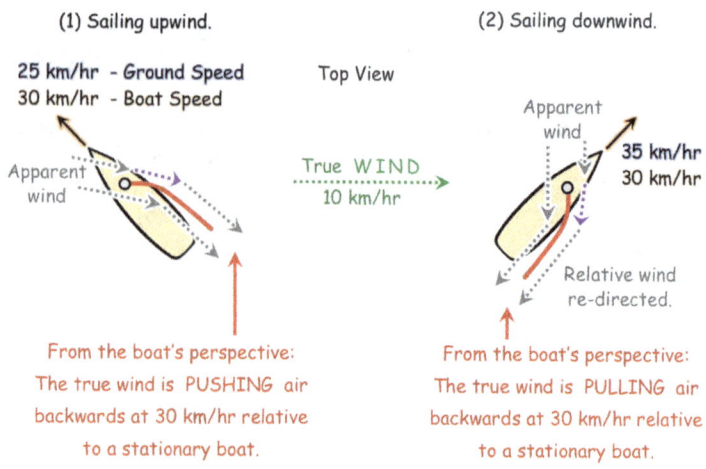

Fig. 11b-i. Relative movement of the wind and boat - 1

From the perspective of the boat sailing downwind, the true wind is perceived as travelling in the opposite direction.

This aspect of the argument is confusing at first, as it is counter intuitive. Yet it breaks no laws of physics and fits with what is observed in practice. It is hard to reason that a tailwind is pulling air backwards relative to the boat sailing downwind. The boat is outrunning its tailwind and accelerating further away. It is the reverse of sailing into the wind faster than the wind.

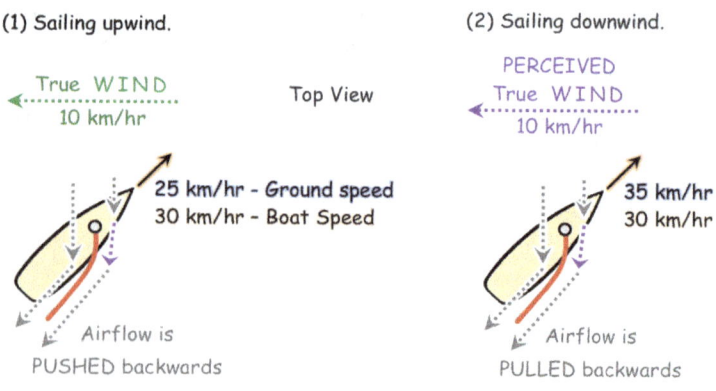

Fig. 11b-ii. Relative movement of the wind and boat - 2

The different relative ground speeds and directions observed on the two different boats in this example are largely irrelevant to the forces generated by the sail. This dynamic is illustrated by boats sailing close hauled in different directions. See Fig. 11b-iii.

Fig. 11b-iii. Apparent wind sailing in opposite directions on a close haul [46]

- **The transfer of momentum from wind to the sail**

According to Newtonian mechanics, a force is created when the sail redirects the relative airflow backwards against the undisturbed true wind, creating turbulence. The turbulence provides something for the re-directed wind to push against to generate the equal and opposite forward force. This process slows down the true wind, extracting momentum from it. The momentum is transferred to the boat. See Fig. 11b-iv.

The sail transfers momentum from the wind to the boat by slowing down the true wind, both upwind and downwind. The wind is the only source of power and energy for a sailboat. See Fig. 11b-v.

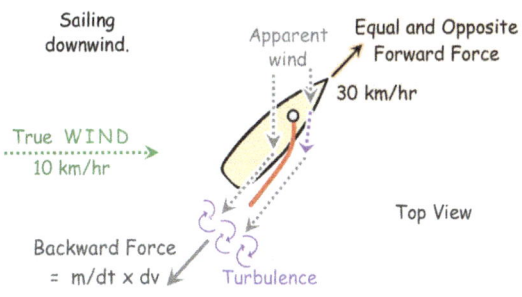

Fig. 11b-iv. Forces acting on a boat sailing downwind

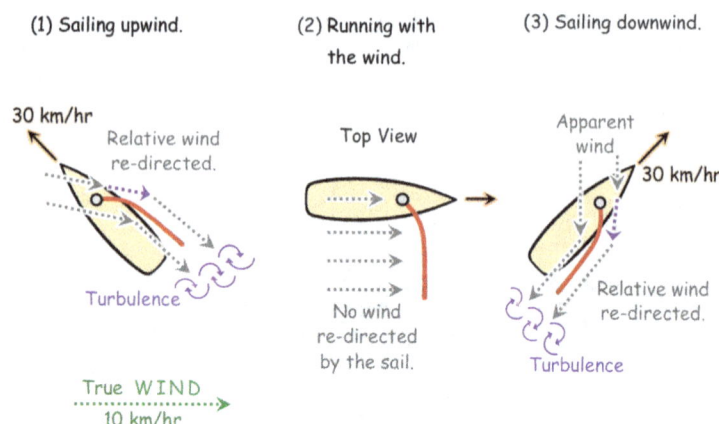

Fig. 11b-v. Transfer of momentum from wind to a sailboat

When the boat is running with the wind at the same speed as the wind, no wind is re-directed by the sail. Therefore, there is little transfer of momentum from the wind to the boat. The force from the wind pushing directly against the boat and the sail is only sufficient to overcome drag from the water. The boat can only travel at the same speed as the true wind.

This transfer of momentum helps explain how a boat can sail downwind faster than the wind. However, this assertion is logical conjecture as it was not possible to test through experimentation. This assertion is the best explanation available that fits with what is observed in practice. It would be useful to test and prove (or disprove) this assertion.

The argument is explained in more detail below.

C. HOW THE APPARENT WIND STAYS IN FRONT OF A BOAT

APPARENT WIND SAILING occurs when a high-performance boat is going much faster than the true wind and the experienced apparent wind is almost always ahead of the sail. [17][18][19]

Apparent wind sailing occurs because the boat speed is so much higher than the true wind, causing the direction of the true wind to be less relevant. The boat's speed (headwind) depends less on whether the boat is sailing upwind or downwind. However, the boat's ground speed is affected by the true wind.

The mathematics for how this dynamic occurs is shown by vector analysis of the true wind, boat speed (headwind), and apparent wind, using example speeds. See Fig. 11c-i.

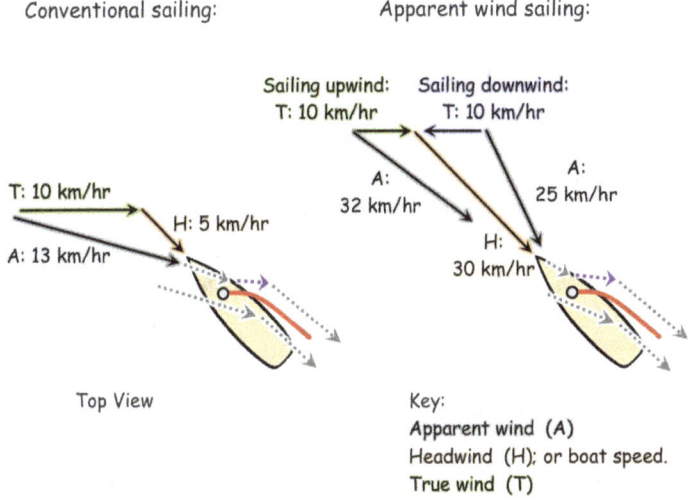

Fig. 11c-i. Example of apparent wind sailing

Competitive sail races provide examples of boats sailing at extremely high speeds on a close haul in almost all directions, almost regardless of the true wind direction. The boat speeds observed are typically multiples of the true wind speed.

In comparison, an open-top car driving down the road at 70 km/hr is always facing into a headwind, as long as the true wind speed is below the car's speed. This dynamic occurs regardless of the car's direction relative to the true wind. Normally, when driving at this speed, the driver has no idea which direction the wind is blowing from. See Fig. 11c-ii.

Fig. 11c-ii. Example of headwind dynamics

Apparent wind sailing into and with the wind

Two sailboats sailing faster than the wind (apparent wind sailing), one sailing downwind and the other sailing into the wind, are exposed to the same true wind, but different apparent winds. Using the example of a 10 km/hr true wind speed, and both boats experiencing a 30 km/hr relative headwind: See Fig. 11c-iii.

- The boat sailing into the wind experiences wind being blown towards it.
- The boat sailing downwind experiences wind being pulled (sucked) towards it; due to Galileo's principle of relative motions, as explained below.

Whether the wind is being pulled or pushed backwards is descriptive. The sailors on each boat cannot tell the difference.

From the perspective of the boat sailing downwind experiencing a headwind, the wind is going backwards relative to the boat, despite the true wind being a tailwind in absolute terms.

Consequently, whether true wind is a tailwind or a headwind, it is mostly irrelevant to a boat apparent wind sailing, both downwind and with the wind. However, the true wind speed and direction is significant for the boat's external references such as its ground speed.

From the reference point of the boat's apparent wind sailing, sailing with the wind is the same as sailing into the wind, but with the apparent wind (relative airflow) blowing in different directions. Either way, both boats experience a headwind.

Sailing into and with the wind use similar diagrams that depict the relative airflows (apparent wind) and forces (Force = m/dt x dv) involved. The diagrams are almost a mirror image.

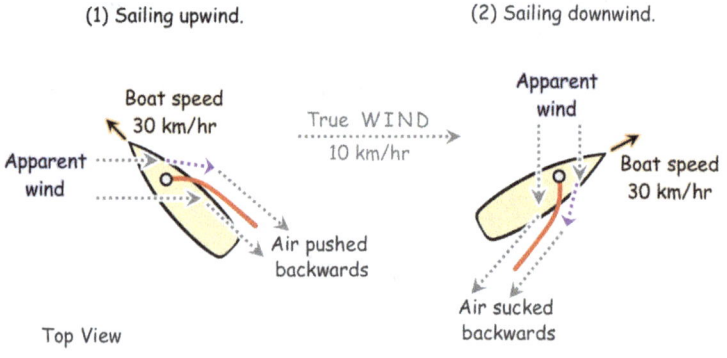

Fig. 11c-iii. Close haul v. broad reach

D. SYMMETRY OF BOAT SPEEDS

THIS ANALYSIS IS consistent with the symmetry of the boat speeds in relation to the wind. The symmetry indicates that the forces involved are universal and constant. The same forces affect how the boat sails upwind and downwind. See Fig. 11d.

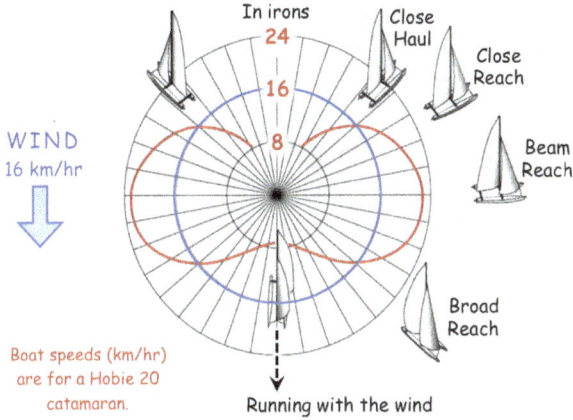

Fig. 11d. Points of sail and example boat speeds. [69]

E. SAILING DOWNWIND FASTER THAN THE WIND

APPARENT WIND SAILING upwind is known to be the same as sailing downwind faster than the wind. However, this enigma has eluded explanation, until now. Unfortunately, the accepted knowledge and understanding of the physics involved has yet to catch up with what is observed in practice.

There is no adequate explanation for the physics of how a boat can sail downwind faster than the wind that has been accepted by physicists. This means that either the known laws of physics are not being interpreted correctly, or that the existing physics tool kit is lacking, and a new approach is needed.

This book argues that the same physics that explains sailing into the wind faster than the wind, also explains sailing downwind faster than the wind. For apparent wind sailing, a close haul into the wind is the same as a broad reach with the wind.

The conundrum is that in apparent wind sailing boats sail downwind several times faster than the wind. It is counter intuitive for a boat to sail downwind faster than the wind when it is powered by the wind. For example, a boat can sail downwind on a broad reach at a boat speed of 30 km/hr with only a 10 km/hr tailwind. See Fig. 11e-i.

There is no accepted and proven explanation for the physics of this conundrum. Many explanations are based on erroneous theories of sailing, vector analysis, and hydrodynamics. Worse, some explanations wrongly compare sailing to the forces created by an airplane wing. Not only is the comparison false, but the theory of airplane flight is also still debated and unresolved. [72]

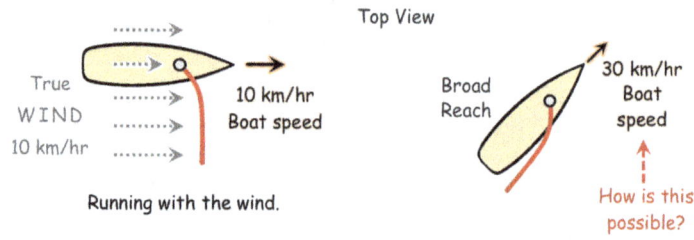

Fig. 11e-i. Sailing downwind faster than the wind

The solution

Sailboats generate a force by extracting momentum and energy from the wind. From the perspective of the boat, the sail achieves this feat by slowing down the apparent wind. Specifically, the sail re-directs the apparent wind, causing turbulence and slows it down. In other words, the sail steals momentum from the apparent wind, which it transfers to the boat.

According to Newtonian mechanics based on the mass flow rate, apparent wind sailing downwind faster than the wind can be achieved in two parts:

- The force (Force = m/dt x dv) generated by the sail depends on the velocity (dv) relative to the boat of the mass of air re-directed each second (m/dt) by the sail. Therefore, the velocity of the apparent wind is only one factor that determines the force generated. See Fig. 11b-iv.

    However, the speed of the apparent wind depends on the speed of the true wind.

    A bigger sail displaces a greater mass of air each second (m/dt), to generate

Downwind faster than the wind

a greater force, which accelerates the boat to a higher speed.

This principle applies equally to downwind and upwind sailing.

In addition, there is a positive feedback loop between the force generated by the sail and the speed of the boat, as described above. The faster the boat goes, the stronger the force generated by the sails.

- All motion is relative, and therefore, apparent wind sailing downwind is essentially the same as apparent wind sailing into the wind, from the perspective of the boat.

It is straightforward to understand how a boat sailing upwind can extract momentum from the wind, because the apparent wind is pushing against the sail.

When sailing downwind faster than the wind, the true wind is pulling (sucking) the air backwards relative to the boat. This allows the sail to redirect the air to generate a force and steal momentum from the true wind.

The true wind needs to be sufficiently strong enough to permit apparent wind sailing.

For example, a boat that is apparent wind sailing at an angle downwind with a boat speed of 30 km/hr in a 10 km/hr tailwind (i.e. true wind) can produce an apparent wind of 25 km/hr. See Fig. 11e-ii.

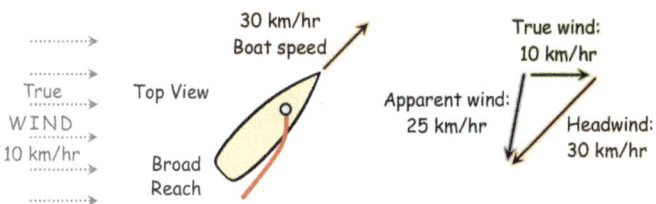

Fig. 11e-ii. Relative headwind sailing downwind

As all motion is relative, the 30 km/hr boat speeds are the same as 30 km/hr headwinds. This means that even though the boat is sailing downwind, it can redirect the relative airflow to generate a force, which is the same as if it were sailing into the wind. A headwind is still a headwind, regardless of whether the boat is sailing into or with the wind.

In contrast, this process and apparent wind sailing is not possible when a true wind is not present. For example, a boat that is accelerated to 30 km/hr by a motor cannot cut the motor and start apparent wind sailing, as there is true wind to slow down and no momentum to steal from the air.

Nicholas Landell-Mills

The transition to downwind sailing

The transition from running with the wind to apparent wind sailing downwind using example speeds is described as follows: See Fig. 11e-iii.

- Drag from the water means running with the wind produces a speed below that of the wind.

- Shifting to a broad reach alters the AOA of the apparent wind, which exposes the leeward side of the sail to the apparent wind. This change boosts the mass of air re-directed backwards each second (m/dt), which is helped by the Coanda effect. In turn, this increases the force generated (Force = m/dt x dv) by the sail, causing the boat to accelerate.

- As the boat gains speed, the AOA of the apparent wind moves towards the front of the boat, causing the boat to transition to apparent wind sailing. The physics allow the boat to exceed the speed of the wind.

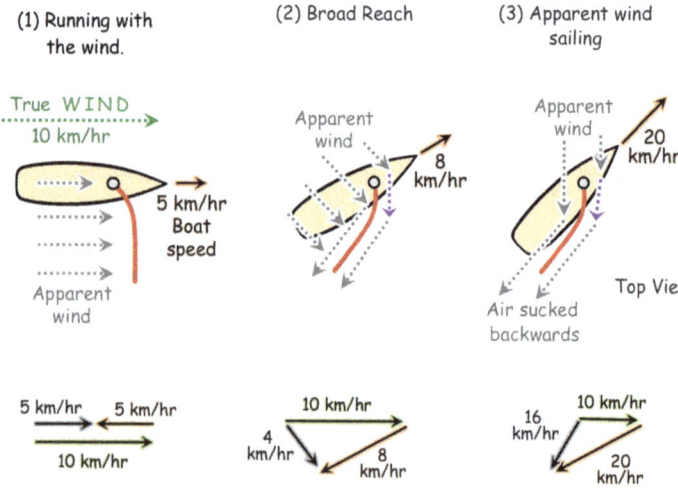

Fig. 11e-iii. Downwind sailing with example speeds

XII

BLACKBIRD LAND YACHT ENIGMA

A. DDWFTTW

THE NEWTONIAN AND Galilean approaches used to explain sailing can be applied to other areas, such as wind turbine propellers, which re-direct a relative airflow to generate a force.

For example, this approach can explain how a propeller driven vehicle, such as the Blackbird land yacht, can travel dead downwind faster than the wind (DDWFTTW). In 2010, the Blackbird vehicle was recorded achieving a speed 2.8 times higher than its tailwind. To date, there is no accepted explanation for this event, which has attracted a lot of attention on physics forums, YouTube, and elsewhere. [26] [27] See Fig. 12a.

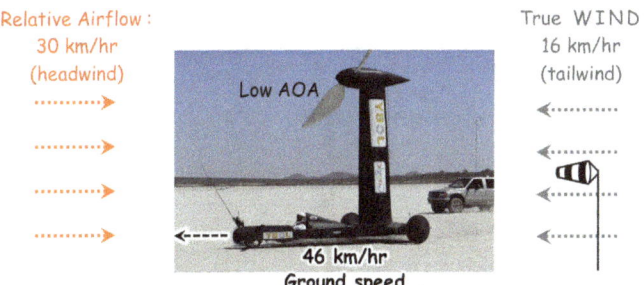

Fig. 12a. Blackbird land yacht [49]

B. THE SOLUTION

THE SOLUTION TO the conundrum is that the contraption is essentially a wind turbine on wheels. The propeller blades have a low AOA to the relative wind and direction of travel.

The twist in each of the two propeller blades either side of a hub, means the blades have a positive AOA to the relative airflow. This aspect means the blades function in a similar manner to two sails aligned with a positive AOA to the apparent wind. See Fig. 12b-i.

Once the vehicle is moving faster than its tailwind (true wind), the relative airflow switches from a tailwind to a headwind. At this point the propeller functions as a wind turbine blade. It generates thrust by re-directing the relative airflow (headwind) backwards, which is similar to how a sail generates a force. The thrust produced turns the blades and pushes the vehicle forward. At this point, the true wind is pulling (sucking) air backwards relative to the vehicle. See Fig. 12b-i.

The propeller blades re-direct relative airflow, which push backwards against the tailwind (true wind) to generate a forward force. This action slows down the tailwind (true wind) and extracts momentum from it, which is transferred to the vehicle.

For example, when travelling at 46 km/hr downwind faster than the wind, the vehicle experiences a 30 km/hr headwind, despite the 16 km/hr tailwind. As all motion is relative, the tailwind is pulling the air backwards relative to the vehicle's movement downwind. The tailwind is not pushing the vehicle forward at this point. This process is almost a mirror image of the vehicle travelling into the wind to generate a force from the propeller. See Fig. 12b-ii.

The physics of the Blackbird vehicle (DDWFTTW) are similar to a boat sailing downwind faster than the wind.

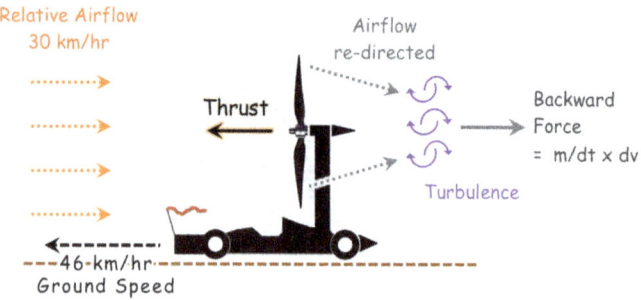

Fig. 12b-i. Relative airflow on the Blackbird vehicle

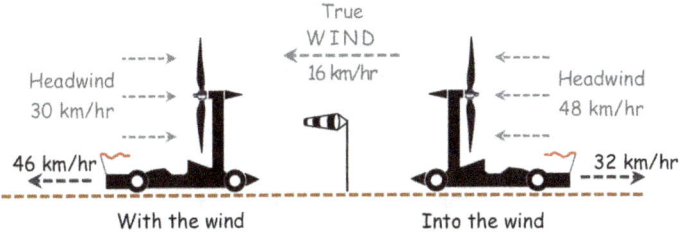

Fig. 12b-ii. Blackbird travelling with and into the wind

C. UPWIND VS. DOWNWIND

THE BLACKBIRD VEHICLE is also reported to have travelled directly upwind at about twice the speed of the wind. Given that the vehicle can travel with the wind with a ground speed almost three times the wind speed; then the difference in ground speeds between travelling into and with the wind is the wind speed. See Fig. 12b-ii

However, in these circumstances the vehicle experiences a much higher headwind when travelling into the wind, as compared to when travelling with the wind. See Fig. 12b-ii

This point is key because it means that the forces involved work equally in both directions. Therefore, there is nothing special or different in the physics between the vehicles moving with or into the wind.

Consequently, similar to a boat apparent wind sailing, whether the true wind is a tailwind or a headwind is mostly irrelevant to everything except the external references, such as the vehicle's ground speed.

D. THRUST = M/DT X DV

THIS ANALYSIS DESCRIBES the forces created by a wind turbine propeller blade turning relatively slowly in a stable rotation through static air with a positive angle of attack (AOA) to a headwind. According to Newtonian mechanics based on the mass flow rate, the airflows and forces are similar to those of a sail. The propeller blade turns too slowly to push air in the desired direction.

The propeller blade re-directs a mass of air each second (m/dt), which it accelerates to a velocity (dv) down and backwards relative to the blade. This action creates a down and backwards force (Force BACK = ma = m/dt x dv). The airflow is re-directed to interfere with the undisturbed wind, creating turbulence, which provides something for the re-directed airflow to push against.

The underside of the blade physically pushes the lower airflow (headwind) down and backwards relative to the blade. The topside of the blade re-directs the upper airflow back and downwards relative to the blade, which is helped by the Coanda effect.

The reaction generates an equal and opposite thrust (Force BACK = Thrust), which turns the propeller blade and creates a forward force. The propeller re-directs air backwards, causing it to be pushed forwards. See Fig. 12d-i and 12d-ii.

The physics of a wind turbine propeller blade re-directing a relative airflow (headwind) is the same as that of a sail, as well as an albatross and a glider soaring. This explanation is different from how an airplane wing and airplane propeller create a force, which is by pushing air backwards.

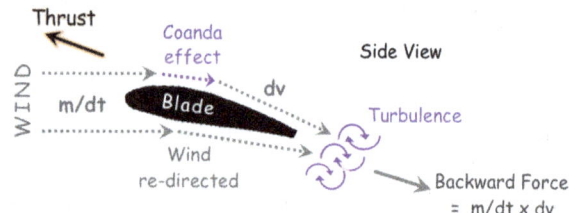

Fig. 12d-i. Newtonian forces acting on a blade

Fig. 12d-ii. Wind Turbine propeller blades with low AOA [50][51][52]

PART 4

FALSE THEORIES OF SAILING

XIII
FALSE THEORIES OF SAILING

A. THE PHYSICS OF SAILING IS UNRESOLVED

THERE IS NO one accepted equation or theory that explains the forces generated sailing. The explanation for the physics of how boats sail into the wind is unresolved because there is no conclusive experiment on an actual boat in realistic conditions that can prove any current theory to be true. More precisely, it is debated exactly how force, momentum, and energy are transferred from the wind to the sail to generate forward motion. The preferred explanations of the forces created by a sail involve complex mathematics and convoluted explanations. See Fig 13a.

The current preferred theories fail to explain sailing because they are based on a few erroneous and unproven assumptions that include: See Fig. 13a.

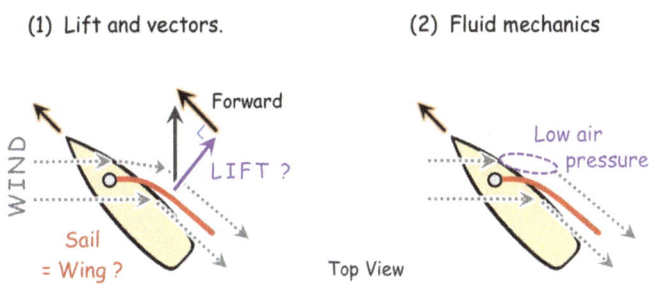

Fig. 13a. False theories of sailing

- Fluid mechanics (hydrodynamics) and Navier-Stokes equations can explain the forces created by a sail.
- A sail creates a force by the wind pushing directly against it, as compared to the sail re-directing the airflow to push against the undisturbed wind.
- A sail has a similar design, shape, and functions to an airplane wing. Therefore, a sail produces similar airflows and forces as compared to lift on a wing. In particular, the sail generates lift perpendicular to the sail. One obvious problem with this argument is that experts have not proven how a wing generates lift. [72]
- Vector-based solutions attempt to explain how the lift generated perpendicular to the sail is channeled into a forward force due to the interaction of the hull, keel, and sail, which allows a boat to sail into the wind.

However, analysis refutes these assumptions by showing that fluid mechanics fails to explain the forces on a sail; a sail is not a wing; and the main force created is not perpendicular to the sail. Past research failed to explain sailing downwind faster than the wind because they relied on these false assumptions. These erroneous assumptions are examined in more detail below.

XIV
FLUID MECHANICS

A. FLUID MECHANICS (HYDRODYNAMICS) AND SAILING

ACADEMICS TEND TO promote theories to explain the physics of sailing into the wind based on hydrodynamics (fluid flow), involving complex mathematical equations (e.g. Navier-Stokes). These theories are devoid of empirical evidence and proof, and are frequently presented without any example calculations. Worse, many academics mistake mathematical proof or computer simulations, such as Computational Fluid Dynamics (CFD), for scientific proof.

According to fluid mechanics, low air pressure created by a fast-moving airflow at the front of the leeward side of the sail is the primary mechanism that pulls (sucks) the boat forward. The problems with this logic presented include:

- There is no experimental proof that fluid mechanics or low air pressure can explain the forces created by a sail.
- It is unclear what this low air pressure pulling against, in order to pull the boat forwards. Perhaps the atmosphere?
- For low air pressure to pull (suck) a heavy boat forward through the water, it must have a large and negative value compared to atmospheric pressure. This is impossible under the circumstances of an open atmosphere, as opposed to a closed environment.
- There is no evidence that even a vacuum (zero air pressure) in front of the sail would be sufficient to pull a heavy boat forward through the water.
- Even if extremely low air existed at the front of the sail, it would pull the air

in front of the boat backwards. The atmosphere provides a lower resistance as compared to pulling a heavy boat forward through the water.

- The telltales on the leeward side of the sail are pushed backwards in the same direction of the re-directed wind, except in the case of turbulent airflow. If low pressure pulled the boat forward, then the telltales on the leeward side of the sail would point forward in the direction of travel, not backwards as observed in practice. See Fig. 14a.

- If debris in the wind passes in front of the sail, it is not sucked forwards. It is blown backwards in practice, which indicates a backward force, not a force pulling the boat forward.

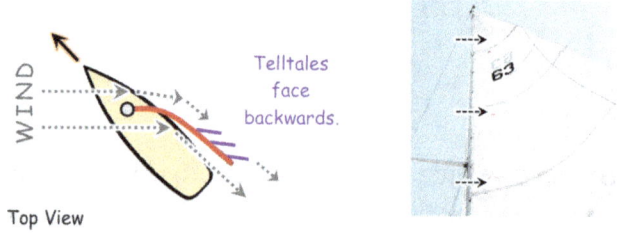

Fig. 14a. Telltales face backwards [40]

The numerous other problems with the argument presented by fluid mechanics include:

- There is no agreed upon equation that describes the force created by a sail. Even the Navier-Stokes equations have numerous variations.

- Almost no animal or object propels itself forward by low air pressure in front of it. Objects and animals move primarily by first pushing backwards, which provides an equal and opposite forward force, as explained by Newtonian mechanics.

- The argument is excessively complicated, creating the risk for error.

- There is no consensus that fluid mechanics explains sailing. Academics have criticized utilizing fluid mechanics and Bernoulli's principles to explain lift as far back as 1972. [5]

- The arguments for fluid mechanics rely on the abstract and unproven Navier-Stokes equations. The problems with these equations are described below.

- It is assumed that fluid flow can be used to describe the resultant lift forces according to fluid mechanics, rather than Newtonian mechanics. There is no evidence that this assumption is correct.
- Advocates of fluid mechanics highlight that fluid mechanics is based on or derived from Newtonian mechanics. However, this logic is inconsistent with the result that the forces created by fluid mechanics pull objects upwards (airplane wings) or forwards (sailboats), rather than pushing them up, which would be consistent with the principles of Newtonian mechanics.
- It is illogical and inconsistent to explain thrust, drag and weight on an airplane or sailboat using Newtonian mechanics. And to then use fluid mechanics to explain lift on a wing or sail.
- NASA uses Newtonian mechanics to explain how kites are pushed upwards. "When the wind hits the front of the kite, the wind is deflected downward, and there is a force in the opposite direction, which pushes the kite upward. This action depicts Newton's Third Law of Motion, ..." But then NASA also promotes fluid mechanics as a solution to explain sailing. [1] See: NASA GRC website, Principles of Flight; for grades K-4. It is inconsistent to explain the force created by a sail from a wind using fluid mechanics, but explain how kites generate a force from a wind using Newtonian mechanics.

Advocates of fluid mechanics frequently defend the use of fluid flow to explain sailing based on the assumption that fluid flow can explain lift by airplane wings, stating that sails are similar to wings. This is a logical fallacy, as fluid mechanics also fails to explain lift by airplane wings. [72]

B. NAVIER-STOKES EQUATIONS [74]

FLUID DYNAMICS INVOLVES complex mathematical equations to explain lift. For example, the prevailing approaches use Navier-Stokes (NS) equations to calculate lift, which involve at least three inter-related simultaneous equations. See Fig. 14b.

The problems using NS equations to explain the lift forces generated by a sail (and airplane wing) include:

- There is no general theory or one NS equation that is applied universally to calculate the generated forces by fluid flow. NS equations have numerous conditions. The theories on fluid flow (e.g. Bernoulli, Kutta effect, Navier-Stokes, etc.) apply to very specific conditions such as steady flow, inviscid flow, incompressible flow, ... Different equations are required for different fluid conditions.

Navier–Stokes Equations
3 – dimensional – unsteady

Glenn Research Center

Coordinates: (x,y,z)
Velocity Components: (u,v,w)
Time: t Pressure: p
Density: ρ Stress: τ
Total Energy: Et
Heat Flux: q
Reynolds Number: Re
Prandtl Number: Pr

Continuity:
$$\frac{\partial \rho}{\partial t} + \frac{\partial (\rho u)}{\partial x} + \frac{\partial (\rho v)}{\partial y} + \frac{\partial (\rho w)}{\partial z} = 0$$

X – Momentum:
$$\frac{\partial (\rho u)}{\partial t} + \frac{\partial (\rho u^2)}{\partial x} + \frac{\partial (\rho uv)}{\partial y} + \frac{\partial (\rho uw)}{\partial z} = -\frac{\partial p}{\partial x} + \frac{1}{Re_r}\left[\frac{\partial \tau_{xx}}{\partial x} + \frac{\partial \tau_{xy}}{\partial y} + \frac{\partial \tau_{xz}}{\partial z}\right]$$

Y – Momentum:
$$\frac{\partial (\rho v)}{\partial t} + \frac{\partial (\rho uv)}{\partial x} + \frac{\partial (\rho v^2)}{\partial y} + \frac{\partial (\rho vw)}{\partial z} = -\frac{\partial p}{\partial y} + \frac{1}{Re_r}\left[\frac{\partial \tau_{xy}}{\partial x} + \frac{\partial \tau_{yy}}{\partial y} + \frac{\partial \tau_{yz}}{\partial z}\right]$$

Z – Momentum:
$$\frac{\partial (\rho w)}{\partial t} + \frac{\partial (\rho uw)}{\partial x} + \frac{\partial (\rho vw)}{\partial y} + \frac{\partial (\rho w^2)}{\partial z} = -\frac{\partial p}{\partial z} + \frac{1}{Re_r}\left[\frac{\partial \tau_{xz}}{\partial x} + \frac{\partial \tau_{yz}}{\partial y} + \frac{\partial \tau_{zz}}{\partial z}\right]$$

Energy:
$$\frac{\partial (E_t)}{\partial t} + \frac{\partial (uE_t)}{\partial x} + \frac{\partial (vE_t)}{\partial y} + \frac{\partial (wE_t)}{\partial z} = -\frac{\partial (up)}{\partial x} - \frac{\partial (vp)}{\partial y} - \frac{\partial (wp)}{\partial z} - \frac{1}{Re_r Pr_r}\left[\frac{\partial q_x}{\partial x} + \frac{\partial q_y}{\partial y} + \frac{\partial q_z}{\partial z}\right]$$
$$+ \frac{1}{Re_r}\left[\frac{\partial}{\partial x}(u\tau_{xx} + v\tau_{xy} + w\tau_{xz}) + \frac{\partial}{\partial y}(u\tau_{xy} + v\tau_{yy} + w\tau_{yz}) + \frac{\partial}{\partial z}(u\tau_{xz} + v\tau_{yz} + w\tau_{zz})\right]$$

Fig. 14b. Navier-Stokes equations [1]

- NS equations are unproven by direct evidence, such as an experiment on a real airplane. Many people mistake mathematical proof of NS equations for scientific proof.

- There is no one accepted NS equation. Academics regularly propose adjustments to the different variations of NS equations.

- NS equations are abstract and difficult to correlate changes in air pressure, viscosity, and friction to the change in the generated forces. NS equations fail to provide an understandable or straightforward explanation for how the mathematical models relate to what is observed in practice.

- It is unclear how viscosity and friction contribute to lift, as these relate to the internal dynamics of the air and not the external forces generated.

- NS equations have a weak theoretical basis. It is uncertain that NS equations are built on a sound theoretical basis. [8][9]

- NS equations are concerned with the internal relationships of a fluid. This leaves experts to speculate how these internal relationships affect lift. NS equations fail to adequately explain and predict turbulence and the physics of an airplane stall. [8][9]

- NS equations have flawed logic. NS equations rely on low air pressure, air viscosity, and friction to create a force. The problems of relying on low air pressure to suck a heavy sailboat through the water are described above.

- NS equations ignore airflows on the windward side of the sail. Only the leeward airflows are assumed to contribute towards the forces generated. This is unrealistic as these windward airflows pushing against the sail create a force.
- NS equations fail to explain aspects of sailing; such as how a boat can sail faster then the wind, both upwind and downwind.
- NS equations are excessively complex. NS equations are too difficult to solve analytically. They are a non-linear set of up to six inter-related, partial differential equations that must be solved simultaneously. The same dependent variables appear in multiple equations. Complexity creates the risk of error in the model and any calculation.
- General solutions do not exist for NS equations, only a few precise solutions. Due to many inter-related variables, it is almost impossible for NS equations to accurately predict how the different variables change with lift. This is also called the existence and smoothness problem. The Clay Mathematics Institute identified NS equations as a Millennial problem in 2000. A $1 million award was offered for proof that they work, or even to provide progress to understanding them. [14]
- NS equations are not used by sailors and rarely used by boat and sail manufacturers to assess the forces generated by a sail.
- NS equations focus on the fluid flow and ignore the vehicle (ie. sailboat or aircraft). It is assumed that the momentum of a sailboat does not affect the generated forces.

Given the long list of problems, limitations, and complications with the NS equations, it is amazing that anyone uses them to calculate the forces generated by a sail. Many of these problems are not restricted to NS equations and can apply to other theories based on fluid mechanics.

XV
FALSE
VECTOR BASED SOLUTIONS

A. FALSE VECTOR-BASED EXPLANATIONS OF SAILING

THE FORCES ACTING on a sail and boat can be described using vectors. See Fig 15a-i.

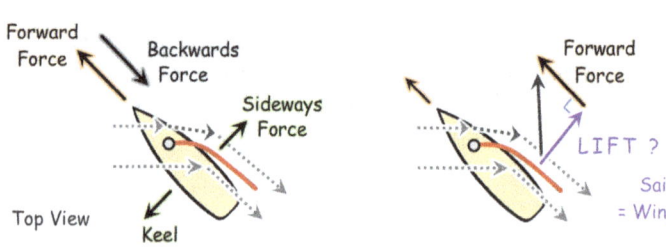

Fig. 15a-i. Vector solutions for a sail

The false vector-based arguments explanations rely mostly on fluid mechanics or aero-hydrodynamic forces, to explain the lift force generated perpendicular to a sail. Often, this force is also mistaken for how an airplane wing generates lift. Overly complex mathematics and convoluted arguments are then used to erroneously explain how a forward force arises perpendicular to the lift force, via mechanical interactions between the wind, sail, hull, keel, and water. [2] [3] [4] [10] [11] [12] [13] See Fig. 15a-ii.

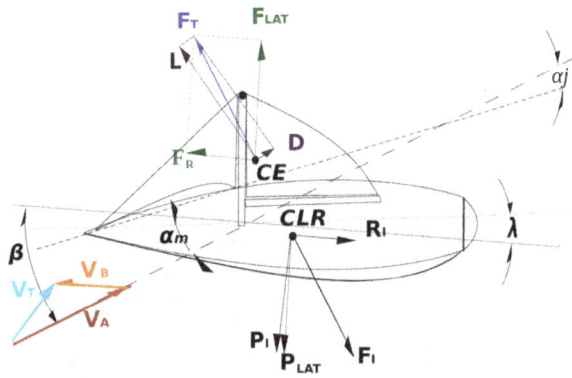

Fig. 15a-ii. Convoluted and complex explanation of sailing [70]

B. PROBLEMS WITH THE FALSE VECTOR-BASED APPROACH

FALSE VECTOR-BASED EXPLANATIONS of sailing into the wind have some fundamental problems that invalidate them. The problems with this approach include:

- This approach is not observed in practice and has not been proven by experimentation on a sailboat.

- There are many different versions of vector-based arguments. There is no one agreed upon explanation of the vectors involved and the physics of how they function to create a forward force that pushes a boat through the water. Worse, vector-based arguments are typically presented without any example calculations, making it difficult to relate to what is observed in practice

- Only the airflow immediately on the sail is considered important. The sail re-directing the apparent wind backwards towards the stern of the boat is ignored.

- The sail acts more like a kite than a sail. The purpose of the sail is to catch air, not to re-direct the relative airflows. Airflows around the sail are ignored and the Coanda effect is not considered to be significant.

- This approach does not specify the equation used to calculate the force generated, nor how efficiently it is channeled upwind.

- Some vector-based approaches rely on the boat's keel or hull channeling the force upwind, without pushing any water downwind. However, catamarans easily sail upwind without a keel. Therefore, any theory of sailing that relies on a keel is false. In fact, catamarans with hydrofoils can sail even faster than boats with one hull, as evidenced by the high-performance boats in the America's Cup. See Fig. 15b.

Fig. 15b. Catamarans sail into the wind without a keel [53][54]

Catamarans rely on two hulls to provide stability and to prevent them from tipping over. This fact provides evidence that the direct force on the sail is only a sideways force, and does not propel the boat forwards.

- This approach does not explain how a boat can sail faster than true wind upwind or downwind. This is partly because these theories are based on the wind pushing directly against the sail to create a force. This limits the force generated by the sail to the speed of true wind.

- Some vector-based approaches used complicated mathematics and rely on assumptions that do not correlate with the physical world.

- Some vector-based theories wrongly assume the sail generates a force like that of an airplane generating lift.

C. A PROBLEM WITH SPLITTING FORCES INTO VECTORS

SOME FALSE VECTOR-BASED arguments rely on a force (Lift) being created by the sail perpendicular to the airflow. Then another force is created perpendicular to this first force (Lift). See Fig 15a and Fig. 15c-i.

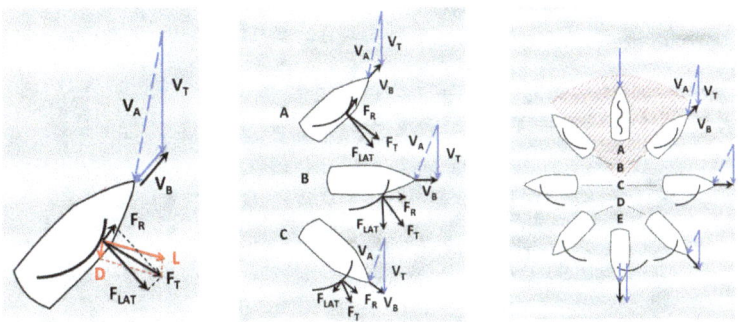

Fig. 15c-i. Vector forces acting on a sailboat [55][56][57]

This approach is very different from the vectors used by Newtonian mechanics to calculate a force based on the sail re-directing apparent wind.

Some vector-based arguments assume that the forces acting on a sail can be split and analyzed according to the mathematical rules of right-angled triangles. However, this is false, as some directional forces may only act in the one direction and cannot be split in two components (or tangents) based on the rules for a right-angled triangle.

In the case of a motorboat travelling in a straight line, the motorboat has a unidirectional force acting on it from the engine thrust. It does not apply any force laterally. It would be false to split the engine thrust according to the rules of a right-angled triangle, except for descriptive purposes. The motorboat's hull is not applying a significant sideways force on the water. See Fig. 15c-ii.

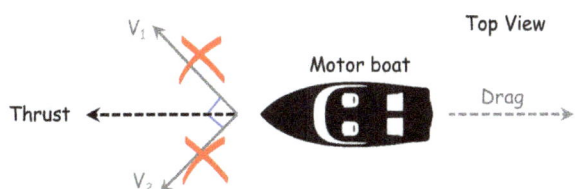

Fig. 15c-ii. Forces acting on a motorboat

A boat and a sail do not exert significant forces perpendicular to the sail or hull as claimed by the false vector-based arguments. This aspect is explained in more detail below comparing kitesurfing to sailing. This aspect is evidence from the wake of a sailboat that shows no water being pushed downwind. See Fig. 15c-iii.

Fig. 15c-iii. Little water is displaced by a boat's hull [63]

D. A VARIETY OF FALSE VECTOR-BASED ARGUMENTS

FALSE VECTOR-BASED ARGUMENTS use a variety of different vectors to explain sailing. There is no one accepted approach. The forces involved in sailing are often incorrectly shown as vectors, as illustrated in Fig. 15d.

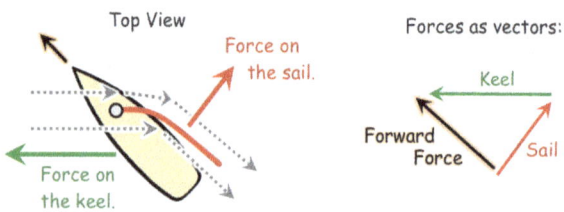

Fig. 15d. Incorrect forces on a boat shown as vectors

E. OTHER FALSE VECTOR-BASED ARGUMENTS.

OTHER VECTOR-BASED ARGUMENTS correctly assert that the re-direction of the apparent wind by the sail creates a force. For example, if the apparent wind of 10 m/s (V_1) were re-directed by the sail to 8 m/s (V_2) at 45° angle, a force would be created due to the change in velocity 'ΔV'. Vectors can be used to illustrate this action. See Fig 15e-i.

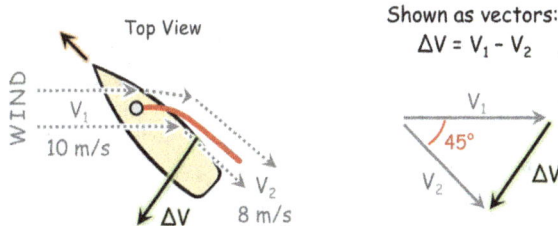

Fig. 15e-i. Forces acting on the sail due to a change in direction of the wind.

The vector-based argument presented above is false because the closer the boat sails into wind (i.e. a lower sail AOA), then the smaller the force 'ΔV', Which is opposite to what is observed in practice. As a boat sails closer into the wind, then the force generated by the sail increases. See Fig 15e-ii.

The force 'ΔV' is the sideways force exerted by the wind on the sail pushing the boat downwind, which causes the boat to tilt. The force is opposed by the keel, if the boat has one, of by the boat tilting (pitching) on to one hull in the case of a catamaran. As the sail's AOA changes the strength and direction of the sideways force (ΔV) also changes correspondingly. See Fig. 15e-ii.

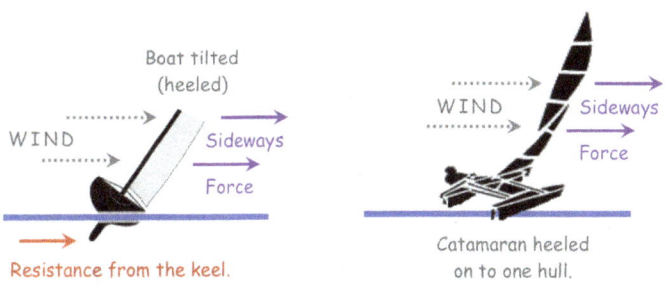

Fig. 15e-ii. Sideways force acting on boats

The momentum of the wind transferred to the boat should be considered per the following example: See Fig 15e-iii.

- The apparent wind is slowed by 2 m/s; from 10 m/s to 8 m/s. The momentum transferred from the wind to the sail can be calculated using the deceleration of the wind and the mass of the air that hits the sail (momentum = mass x velocity).

- As the deceleration of the apparent wind of 2 m/s is limited, the momentum transferred to the sail by the sideways force is also limited.

- The apparent wind re-directed by the sail interacts with the undisturbed apparent wind after exiting the sail, to create turbulence. It is assuming that re-directed wind decelerates from 8 m/s to 0 m/s in the turbulence. This represents four times the momentum transferred to the sail via the sideways force.

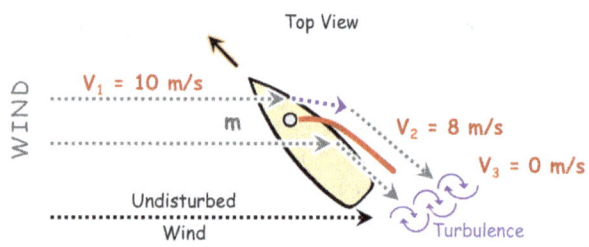

Fig. 15e-iii. Apparent wind velocity sailing

On the other hand, the vector-based arguments take a completely different approach. They assert that the force 'ΔV' creates an equal and opposite force that represents the lift on the sail. In turn, the lift pushes the boat ahead due to the vector-based arguments above. See Fig. 15e-iv.

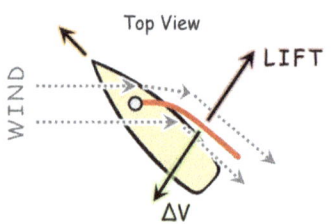

Fig. 15e-iv. Generation of lift on a sail according to vector-based arguments

This assertion by the vector-based arguments is wrong because it confuses lift with the sideways force 'ΔV' on the sail. The force 'ΔV' that is approximately perpendicular to the sail cannot push the boat forward.

XVI
COMPARISON
TO KITESURFING

A. SAILING COMPARED TO KITESURFING

THE KITES USED in kitesurfing have the same basic design and function as compared to a sail on a boat; albeit the shape of a kite is closer to a parapont than a sail on a boat. Kites and sails produce similar airflows, and therefore, similar forces. Newtonian physics based on the mass flow rate can be used to explain how a kitesurfer generates a force, consistent with how boats sail into the wind.

B. THE PHYSICS OF KITESURFING

KITESURFING IS ESSENTIALLY a wakeboard with a paraglider. The kite redirects a mass of air each second (m/dt) from the apparent wind and helped by the Coanda effect, at a velocity (dv) backwards relative to the kite. This redirected airflow pushes against undisturbed apparent wind behind the kite creating turbulence, which provides something to push against. This action generates a backward force (Force = ma = m/dt x dv). The reactive equal and opposite forward force (Force $_{KITE}$) pushes the kite ahead. See Fig. 16a and 16b.

The kite pulls the kitesurfer forwards similar to how a motorboat can pull a person water-skiing or wakeboarding alongside the boat forwards. The wakeboarder uses their board to push water towards the boat, in order to maintain line

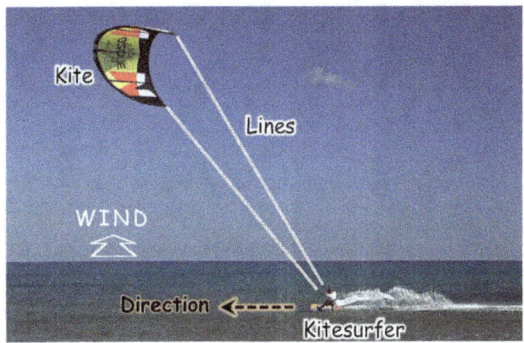

Fig. 16a. Kitesurfing [60]

tension and control their direction. These aspects can be demonstrated by replacing a speedboat with a kite in the image of a person wakeboarding. See Fig. 16c.

C. THE DOWNWIND FORCE

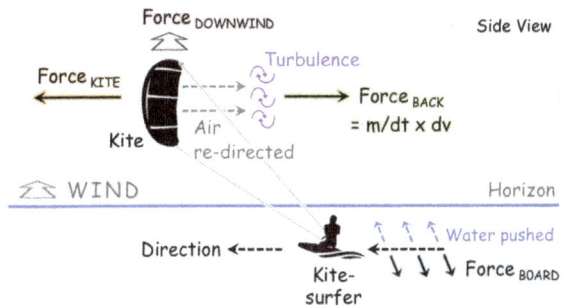

Fig. 16b. Forces acting on a Kitesurfer

THERE ARE TWO separate forces acting on a kite, which include:

- A relatively small downwind force directly against the kite, due to the kite re-directing the wid. This force is the equivalent of the sideways force acting on a sail that causes the boat to tilt. The downwind force keeps the kite open and inflated.

Fig. 16c. Wakeboarding alongside a motorboat [58]

- A larger forward force (Force KITE) that propels the kite ahead, due to the airflow re-directed backwards by the kite.

To avoid being pushed downwind by the downwind force, the kitesurfer leans into the wind and uses the board to push water downwind. The equal and opposite force on the board pushes the kitesurfer upwind. In short, the water provides resistance against the kitesurfer being blown downwind. See Fig. 16d.

When sailing into wind, dynamic is evident because kiteboards only produce a wake on the downwind side, in the direction that the water is being pushed. No wake is evident upwind of the kitesurfer. See Fig. 16e.

In contrast, the sideways force on a sail causes the sailboat to tilt with the wind, which is resisted by the boat's hull and/or keel. However, the hull and keel do not push much water downwind, as evident from images of catamarans and boats sailing into wind. See Fig. 16f and 16g.

This aspect makes sailing a more efficient method of propulsion as compared to kitesurfing.

Fig. 16d. Kitesurfers pushing water downwind [61][62]

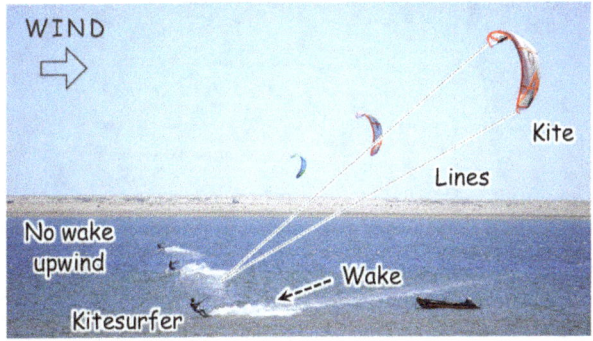

*Fig. 16e. Wake from Kitesurfers sailing into the wind [59]
wind; boat tilting with the wind*

*Fig. 16f. The wake from Catamarans
sailing into the wind [35]
sailing into the wind*

*Fig. 16g. The wake on a boat
sailing into the wind [63]
sailing into the wind*

XVII
A SAIL
IS NOT A WING

A. AIRFLOWS AND FORCES ARE DIFFERENT

THE PHYSICS OF sailing is frequently compared to the physics of how airplane wings generate lift. Both are considered to involve similar relative airflows moving around airfoils (wing / sail) to produce forces that enable movement.

However, a difference between sails and wings is the different transfer of momentum and impetus for the airflows that generate the resultant forces. In particular, a fundamental difference is that a sail re-directs a moving airflow to transfer momentum from the wind to the boat. Whereas a moving wing pushes static air in a desired direction, transferring momentum from the aircraft to the air to generate lift.

A moving airplane wing is horizontally aligned to push static air downwards to create lift against gravity, compared to a sail that re-directs airflow from an apparent wind to create a forward force. See Fig. 17a.

A sail and wing have sufficiently different designs, shapes, and functions to create distinctively different airflows and therefore, forces. This is evident by the following:

- An airplane does not fly efficiently with a wing shaped like a sail. Also vice versa, a boat does not move through the water efficiently with a sail designed like a wing.
- Wings usually travel at a higher speed as compared to sails.

- Wings circulate air behind them around wingtip vortices, whereas, sailboats create airflow turbulence in their wake.

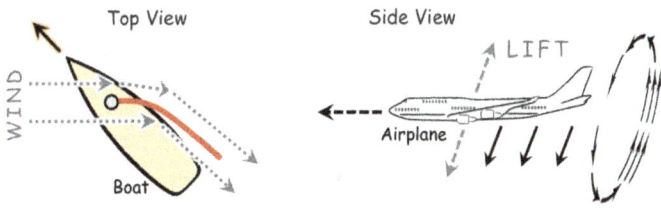

Fig. 17a. Lift generated by a sail and a wing

Exceptions to this assertion include a glider wing that acts like a sail when soaring to re-direct airflows, and high-speed boats that uses rigid or wing-shaped sails (e.g. sailrocket). The optimum wing or sail design and shape changes with function and circumstances (e.g. speed).

The Newtonian physics and equations based on the mass flow rate can be used to explain the lift force generated by a wing and hydrofoil, as well as the forward force generated by a sail (Force = ma = m/dt x dv). However, the application of Newtonian physics is slightly different as a wing and hydrofoil push air down, while a sail re-directs airflow to create a force.

In summary, the difference between a sail and a wing is based less on design and shape, but on function. All motion is relative, but there is a difference between how a wing and a sail create airflows. There is also a difference in the source of the momentum used to a generate force by each object. In addition, static and moving air behaves differently. This logic supports the conclusion that a sail is not a wing.

B. THE THEORY OF LIFT IS UNRESOLVED [72]

THE PHYSICS OF lift is disputed. Currently there is no scientific experiment on a real aircraft in realistic conditions that proves any theory or equation for lift to be true. Therefore, it is wrong to assert that sails generate a force in the same way that wings generate lift.

Fig. 17b. It is unclear how lift is generated by a wing

Experts cannot agree on whether aircraft achieve lift by being pulled upwards according to fluid mechanics or pushed upwards according to Newtonian mechanics; nor can they agree on what role vortices play.

Academics, engineers, aircraft manufacturers, pilots, aviation authorities, and other pundits (e.g. NASA) promote more than 12 diverse theories of lift and new theories are occasionally proposed. This is surprising given airplanes have been flying for over a hundred years.

The media occasionally comment on the ongoing debate about the mysterious, unproven, and unknown causes of lift:

- "Staying Aloft; What Does Keep Them Up There?" in New York Times, 2003. [20]
- "How Do Airplanes Fly?" in Live Science, 2006. [22]
- "The secret to airplane flight. No one really knows." National Newspaper, 2012. [21]
- "There's No One Way to Explain How Flying Works," Wired Magazine, 2018. [23]
- "No One Can Explain Why Planes Stay in the Air." in the Scientific American magazine, 2020. [24]
- "Quest for an Improved Explanation of Lift," in the AIAA journal, 2012. [25];

In summary, if experts cannot agree or prove how a wing creates lift, then it is inconsistent and illogical for experts to argue that how a sail creates lift to generate a forward force is similar to how a wing generates lift.

C. WINGS AND SAILS COMPARED

Wings are symmetrical with a flat underside and are made of solid materials designed to fly at high speeds. Whereas sails are asymmetrical and have a hollow or concave windward side made of a flexible fabric designed to sail at relatively low speeds. See Fig. 17c-i and 17c-ii.

The main similarity between these two objects is the curved topside of a wing and the curved (convex) leeward side of a sail. The purpose of this is to maximize the Coanda effect and airflows on these surfaces. But curvature is not enough.

It would not benefit a sailboat to generate a sideways force perpendicular to the sail similar to the lift on an airplane wing. If a sail did generate force similar to a wing, it would push the boat sideways.

A wing generates lift by transferring momentum and energy from the airplane to the air. Whereas a sail generates forward motion by transferring momentum and energy from the apparent wind to the sailboat.

However, a sail is similar in principle to a glider or albatross wing soaring; which re-direct relative airflow from an apparent wind to create lift.

In addition, a wing (airfoil) and a hydrofoil have similar designs and functions. They generate similar fluid flows and forces, which are somewhat different from those of a sail.

Fig. 17c-i. Sail and wing compared [63][66]

Fig. 17c-ii. Airplane wing cross-sections [64][65]

Newtonian mechanics based on the mass flow rate explains lift from a wing and sailing, but it is applied differently due to the differences in airflows.

In summary, it is wrong to state that the way wings generate lift is the same as how sails propel boats forward, because the airflows and forces are different.

D. WINGS CIRCULATE THE AIR; SAILS DO NOT [73]

AIRPLANES CIRCULATE AIR displaced in flight. The air flown through and displaced down with gravity then pushes air up elsewhere to replace the space vacated by the air displaced down. This dynamic of circulating air does not apply to sailboats, as sails do not circulate air against gravity in the same manner. See Fig. 17d-i.

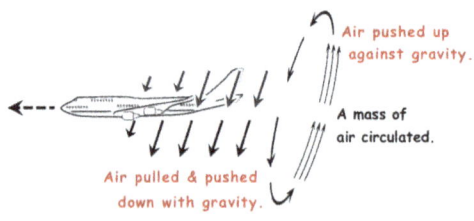

Fig. 17d-i. Airplane wings circulate air in flight

Airliners flying through clouds provide clear evidence of air being circulated in flight. The wings push and pull the air flown through downwards. This action then circulates the air laterally to each side of the wings. There are two separate and identical circular airflows on each side of the airplane, centered on wing-tip vortices.

The air being circulated looks like two separate swirls of counter rotating air. The significant part is not just the vortices, but the large mass of air that is

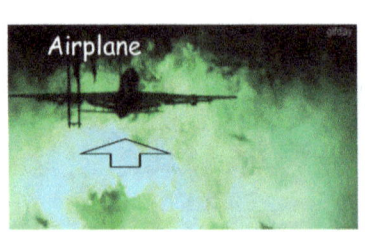
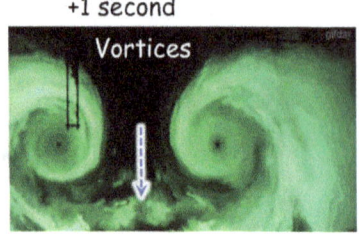

Fig. 17d-ii. Vortices created by airliners 1 [1]

circulated around the vortices. The circulated air covers a much wider area than the relatively small wingtip vortices. See Fig. 17d-ii.

Sails do not produce these airflow patterns seen behind airplanes in flight.

E. WINGS: LIFT = M/DT X DV [73]

FOR REFERENCE, A summary of how an airplane generates lift according to Newtonian mechanics based on the mass flow rate is provided below.

In stable cruise flight, wings with a positive angle of attack (AOA) fly through a mass of air each second (m/dt) and accelerate this air to a velocity (dv) downwards and slightly forwards. This action creates a downward force due to Newton's 2nd Law of Motion (Force = ma = m/dt x dv).

The equal and opposite upward force generated pushes the airplane up and provides lift. Lift is the vertical component of the upward force. Simply put, air goes down and the airplane goes up. See Fig. 17e-i.

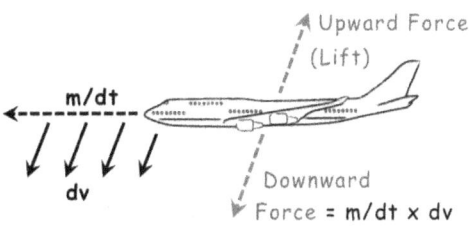

Fig. 17e-i. Newtonian forces on an airplane

Additional considerations include the following:
- Momentum is transferred from the airplane to the air to generate lift via the wing.
- This action only relates to the wings. It does not include the effects on lift and drag from the tail or fuselage.
- There are two separate airflows from a wing. One airflow above and one below the wing.
- 'dv' is provided as a single number. It is a weighted average of each 'dv' for the two different airflows, below and above the wing.

- The airplane's engines can be angled slightly upwards in cruise flight, which contribute towards vertical lift.
- As the wings displace air downwards and slightly forward, the downward and upward forces are at a slight angle to the vertical direction. Lift is the vertical component of the upward force. See Fig. 17e-ii.

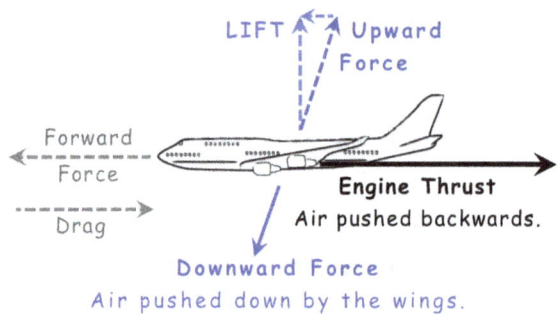

Fig. 17e-ii. Key forces on an airplane (Newtonian mechanics)

F. WING–SAILS DESIGNS

ATTEMPTS HAVE BEEN made to use wing designs as sails. These have been largely unsuccessful, which further confirms that sails and wings are very different.

A hybrid of sails and airplane wings are being developed, called wing-sails, which replace the conventional main sail with a sail shaped as a hollowed-out wing. The main change is to alter sail curvature on the windward side. See Fig. 17f-i.

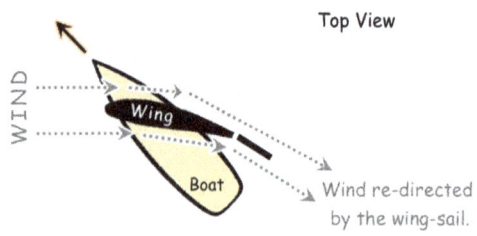

Fig. 17f-i. Wing-sail design on a boat

In addition, the wing-sail is often made of the same material as before, or rigid. Some of the new wing-sails designs on sailboats remove the jib, and rely only on the main sail to generate a force.

Other designs split the sail into two parts. The forward part is attached to the mast that resembles a wing, while the rear part resembles the standard sail design. This allows the different parts of the sail to be adjusted separately. In turn, this provides additional airflow control via altering the sail AOA and the sail shape. A gap between the two parts of the sail also affects airflow.

The physics of how a force is generated by a sail from an apparent wind does not change. Wing-sails represent a potential efficiency improvement, by enhancing airflow management to produce an optimum force from the wind. They do not alter the physics involved in sailing.

There are several problems that question any assertion that wing-sails are better than conventional sails:

- No experimental evidence proves that a wing-sail design alone is more efficient than a conventional sail. A wing-sail is unlikely to be better when sailing with the wind.
- A wing-sail is symmetrical, which is likely to be less efficient than a conventional sail.
- A wing of a motorized airplane and a sail have unique functions and affect airflows differently. A sail has a similar function to an albatross or glider soaring.

Some of these issues are described in more detail below.

Problem: Symmetrical wing-sails

Most conventional wings are asymmetrical, being flat on the underside and curved on the topside (which maximizes the Coanda effect, similar to the leeward side of a sail.

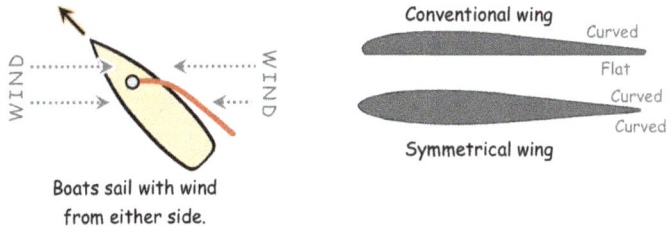

Fig. 17f-ii. Wing-sails need to be symmetrical

However, a wing-sail is symmetrical, similar to the wing of an aerobatic airplane. This means that the windward side of the sail has the same curved surface facing the wind as the leeward side, which is less efficient that a conventional sail. This problem removes much of the benefit of designing a sail like a wing. See Fig. 17f-ii.

SUMMARY: WING-SAILS

A WING-SAIL CANNOT generate lift from apparent wind in the same way as the wing of a powered airplane flying through static air. However, a sail can generate a forward force in the same way the wings of an albatross or a glider performs while soaring.

PART 5

CONCLUDING REMARKS

XVIII

DISCUSSION

A. NEWTONIAN MECHANICS EXPLAINS SAILING

NEWTONIAN MECHANICS BASED on mass flow rate, the Coanda effect and the principle of Galilean relativity provides a new explanation of the physics of sailing and what is observed in practice. This explanation provides a straightforward and easily understood way to describe how boats sail, which is not currently available. It avoids overly complex equations, like Navier-Stokes equations, based on fluid mechanics. Fluid mechanics explains the airflows involved in sailing, but not the resultant forces.

A correct understanding of the physics of sailing provides new and useful insights, which makes it easier to learn how to sail and provide the basis for improved sail design and racing techniques. This is significant to sailors as the underlying physics of sailing is fundamental to the sport. Currently resources are being wasted pursuing false theories, as well as sub-optimal sailing techniques and sail designs.

For example, a useful insight is that All practical aspects of sailing, such as sail size and shape, can be analyzed in terms of the impact they have on 'm/dt' and 'dv' and thus the generated force (Force = m/dt x dv). Fig. 18a.

Another insight is that the area of the sail is not a key determinant of the force generated by the sail. The wing reach, sail AOA, and sail height (wingspan) are better metrics to measure 'm/dt,' and consequently, the force generated (Force = m/dt x dv).

This approach also provides a method to more accurately calculate the force, kinetic energy and power generated by a sail, using the same parameters of 'm/dt' and 'dv.'

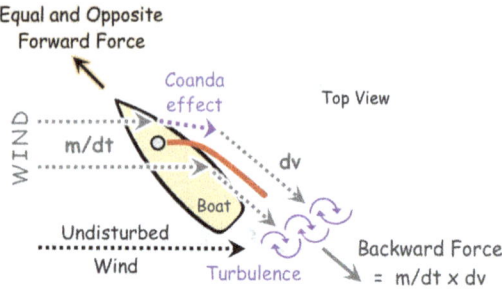

Fig. 18a. Newtonian forces acting on a boat sailing into wind

Specifically, this new approach can explain several nautical conundrums, including:

- How a boat's speed increases when sailing closer into the wind, despite having less sail area exposed to the wind.
- Why boats sail better with multiple sails, instead of one large sail with the same total sail area.
- How a boat can sail both upwind and downwind faster than the wind.

B. HYDROFOILS; LIFT = M/DT X DV [73]

NEWTONIAN MECHANICS CAN be applied to explain the lift generated by a hydrofoil. Hydrofoils and airplane wings (airfoils) have the same basic design, shape and function. Hydrofoils are simply wings (airfoils) that generate lift by pushing water instead of air downwards. See Fig. 18b.

A key difference is that hydrofoils are significantly smaller than wings. This is partly because water is about 1,000 times denser than air. Therefore, according to Newtonian mechanics, to generate the same amount of lift (Lift = m/dt x dv), the hydrofoil needs to displace 1,000 times less volume of water downward, as compared to a wing displacing air downward. Also, boats tend to be a lot lighter than airplanes, as sailboats tend not to have engines, avionics equipment, landing gear…

According to Newtonian mechanics based on the mass flow rate, a hydrofoil has a positive angle of attack (AOA). The hydrofoil moves through a mass of air each second (m/dt), which it accelerates to a velocity (dv) downwards to create a downward force (Force = ma = m/dt x dv).

The reaction generates an equal and opposite upward force that provides lift. Lift is the vertical component of the upward force. Simply put, the hydrofoil pushes water down, causing the boat to be pushed up. See Fig. 18b.

Momentum is transferred from the boat to the water to generate lift via the hydrofoil.

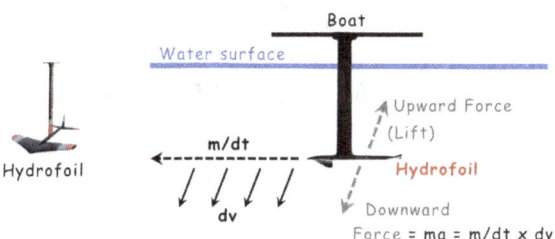

Fig. 18b. Hydrofoil and Newtonian forces acting on a hydrofoil

Both sails and hydrofoils rely on Newtonian physics to explain the forces created based on the mass flow rate. However, an important difference between hydrofoils and sails is that a sail re-directs airflows to create a force against the atmosphere, whereas a hydrofoil pushes water down to create a force against the hydrofoil.

The key benefit of a hydrofoil is to raise the boat out of the water, which reduces the drag from the water when sailing.

C. VORTICES AND WIND SHEAR

THIS ANALYSIS DOES not account for vortices or wind shear created by the sail, except to the extent that these affect 'm/dt' and 'dv'. Contrary to the views of some nautical experts, there is no conclusive evidence that vortices and wind shear are important to the generated forward force. Although these elements can be important to the sail's performance, they are not of primary importance to the forces created to push the boat ahead.

XIX
CONCLUSIONS

IN CONCLUSION, NEWTONIAN mechanics based on mass flow rate and the principle of Galilean relativity provides a new explanation of the physics of sailing and what is observed in practice. This approach offers useful insights. It explains how boats sail downwind and upwind faster than the wind.

XX ADDITIONAL INFORMATION

Corresponding authors: Mr. N. Landell-Mills. Email: nicklm@gmx.com.

Background: The author is British and was born in 1966 in Botswana. He is an independent researcher.

Affiliations: The author is a graduate of the Dept. of Humanities and Social Sciences, The University of Edinburgh, Edinburgh, UK. He was awarded a M.A. degree class 2:1.

Author Contributions: This paper is entirely the work of the author, Mr. Nicholas Landell-Mills.

Disclaimer: The author confirms and states that all data in the manuscript are authentic, there are no conflicts of interest, and all sources of data used in the paper are acknowledged where possible.

Funding: This paper was self-funded by the author.

Acknowledgments: None.

XXI
REFERENCES

[1] NASA, Glenn Research Centre. www.grc.nasa.gov

[2] Wolfgang Püschl, High-speed sailing, Published 8 May 2018, European Physical Society, European Journal of Physics, Volume 39, Number 4

[3] Byron Anderson, The physics of sailing into the wind, Physics Today Magazine, Feb 2008. Volume 61, Issue 2, Page 38, DOI: 10.1063/1.2883908. https://physicstoday.scitation.org/doi/10.1063/1.2883908

[4] Ryan M. Wilson, The Physics of Sailing, JILA and Department of Physics, University of Colorado, USA, February 7, 2010.

[5] NF Smith (1972); Bernoulli and Newton in Fluid Mechanics, the Physics Teacher Journal, (AAPT), volume 10; Published online in 2006 at: https://doi.org/10.1119/1.2352317.

[6] R. Crowley 2015, Conquerors: How Portugal Forged the First Global Empire. Random House Publishing, ISBN-10 : 0812994000; ISBN-13 : 978-0812994001.

[7] M. George, 2017, Te Laa o Lata of Taumako: Gauging the performance of an ancient Polynesian sail, Pacific Traditions Society, DOI: 10.15286/jps.126.4.377-416.

[8] Quanta magazine, Navier-Stokes equations, 26 Feb 2021, https://www.quantamagazine.org/tag/navier-stokes-equations

[9] K. Hartnett; Quanta magazine, Mathematicians Find Wrinkle in Famed Fluid Equations; 21 Dec, 2017; https://www.quantamagazine.org/mathematicians-find-wrinkle-in-famed-fluid-equations-20171221/.

[10] MIT website: HOW A SAIL BOAT SAILS INTO THE WIND; https://web.mit.edu/2.972/www/reports/sail_boat/sail_boat.html

[11] American Institute of Physics. Feb 2008; Physics Today Magazine, Volume 61, Issue 2; 10.1063/1.2883908, https://physicstoday.scitation.org/doi/10.1063/1.2883908

[12] Cristen Conger "How Sailboats Work" 11 March 2008. HowStuffWorks.com. 10 February 2021; < https://adventure.howstuffworks.com/outdoor-activities/water-sports/sailboat.htm >

[13] Indiana University, College Textbooks, Physics, 13 February 2021; https://iu.pressbooks.pub/openstaxcollegephysics/chapter/bernoullis-equation/

[14] Clay Mathematical Institute, Millennium problems, www.claymath.org

[15] J. Sokol; Quanta magazine, Mathematicians Tame Turbulence in Flattened Fluids, 27 June 2018, https://www.quantamagazine.org/mathematicians-tame-turbulence-in-flattened-fluids-20180627/

[16] Bernoulli's Principle; Nov 10, 2007; YouTube channel: phyisfun; https://youtu.be/WDGNcmEOjs4

[17] Bethwaite, Frank (2007). High-performance Sailing. Adlard Coles Nautical. ISBN 978-0-7136-6704-2.

[18] Bethwaite, Frank (2008). Higher performance sailing. London: Adlard Coles Nautical. ISBN 978-1-4729-0131-6. OCLC 854680844.

[19] Jobson, Gary (1990). Championship Tactics: How Anyone Can Sail Faster, Smarter, and Win Races. New York: St. Martin's Press. pp. 323. ISBN 0-312-04278-7.

[20] K. Chang (Dec 9, 2003), Staying Aloft; What Does Keep Them Up There? New York Times. See: www.nytimes.com or https://www.nytimes.com/2003/12/09/news/staying-aloft-what-does-keep-them-up-there.html

[21] R Matthews (Jan 1, 2012), The secret to airplane flight. No one really knows. The National newspaper, UAE.

[22] RR Britt (August 28, 2006), How Do Airplanes Fly? in Live Science.: https://www.livescience.com/7109-planes-fly.html

[23] R. Allain, There's No One Way to Explain How Flying Works; Wired Magazine, 22 Feb 2018, Website: https://www.wired.com/story/theres-no-one-way-to-explain-how-flying-works/

[24] E Regis, No One Can Explain Why Planes Stay in the Air. 1 Feb 2020, Scientific American Magazine. https://www.scientificamerican.com/article/no-one-can-explain-why-planes-stay-in-the-air/

[25] J Hoffren (2012), Quest for an Improved Explanation of Lift, AIAA Journal, Helsinki University of Technology, https://doi.org/10.2514/6.2001-872

[26] K. Barry, Wired Magazine, 6.02.2010, Wind-Powered Car Travels Downwind Faster Than the Wind; www.wired.com/2010/06/down-wind-faster-than-the-wind/

[27] Risking My Life To Settle A Physics Debate; May 29, 2021; YouTube channel: Veritasium; https://youtu.be/jyQwgBAaBag

Images:

[28] Cape Horn Engineering, UK; www.cape-horn-eng.com

[29] Wiki commons – by Pearson Scott Foresman, Sept 2020.

[30] Wiki commons – by Walrasiad in 2010,

[31] Wiki commons – by Rama. Sept 2005.

[32] Image by Ben Kerckx from Pixabay

[33] PublicDomainPictures from Pixabay.

[34] Image by ron-dauphin from unsplash.

[35] Wiki commons – America's_Cup 2015, by Peter Trimming;

[36] Image of Davis Wing licensed from Critical Past; www.criticalpast.com .

[37] Wiki commons – by RAF, June 2014.

[38] Image by Eigil Nybo from Pixabay.

[39] Images of wing in a wind tunnel licensed from Critical Past; www.criticalpast.com

[40] Image of sail by andrew-neel on unsplash.

[41] Image of sail by andrew-neel on unsplash.

[42] Airboat tours with TourBigEasy www.tourbigeasy.com .

[43] Wiki commons – by by Don Ramey Logan, March 2013.

[44] Wiki commons – by Donan.raven; Sept 2013; Oracle team;

[45] Image ahsing888 from Pixabay

[46] Wiki commons – by Peter Trimming; July 2015;

[47] Wiki commons – by Dobromir Slavchev; Aug 2020;

[48] Wiki commons – by U.S. Navy - Feb 2017.

[49] Wiki commons – by Stephen Morris, CC BY-SA 3.0, https://commons.

wikimedia.org/w/index.php?curid=34916718

[50] Wiki commons – by Muzaffar Bukhar, Oct 2012.

[51] Wiki commons – by Phil Hollman, July 2006.

[52] Wiki commons – by Alex DeCiccio, Dec 2016.

[53] Wiki commons – by Nilfanion; Sept 2011.

[54] Wiki commons – by Nilfanion; Sept 2011.

[55] Wiki commons – by HopsonRoad, March 2015.

[56] Wiki commons – by HopsonRoad, March 2015.

[57] Wiki commons – Andrew c, March 2015.

[58] Wiki commons – Wakeboarding Lake Mead, by NPGallery, July 2020.

[59] Wiki commons – Dakhla-Morocco, by Nomadz, March 2009

[60] Wiki commons – by Bвласенко, Sept 2013.

[61] Wiki commons – by Andrzej Otrębski - March 2021.

[62] Wiki commons – by Michal Osmenda, Oct 2006.

[63] Wiki commons – by Don Ramey Logan, April 2016.

[64] Wiki commons – by Robert Frol, March 2011.

[65] Wiki commons – by Smallbones, May 2010.

[66] Wiki commons – by Ad Meskens Sept 2017.

[67] Wiki commons – by Andrew c, Jan 2007.

[68] Wiki commons – by DonSimon - Aug 2019.

[69] Hobie Cat 16 on WordPress.com; Points of Sail; https://hobiecat16.wordpress.com/2010/09/23/points-of-sail-vs-boat-speed/

[70] Wiki commons – by Bcebul; March 2015; Aerodynamic forces on a sail.

[71] Wiki commons – by Tony Hisgett - June 2015.

Unpublished papers by the author:

[72] N Landell-Mills (2019), How airplanes generate lift is disputed (Newton v. Fluid Mechanics). Pre-Print DOI: 10.13140/RG.2.2.34380.36487.

[73] N Landell-Mills (2019), Newton explains lift; Buoyancy explains flight. The physics of how airplanes stay airborne. Pre-Print DOI: 10.13140/RG.2.2.16863.82084.

[74] N Landell-Mills (2020), Why Navier-Stokes equations fail to explain lift; Pre-Print DOI: 10.13140/RG.2.2.10678.52809.

[75] N Landell-Mills (2021), Propeller thrust explained by Newtonian physics; Pre-Print DOI: 10.13140/RG.2.2.17375.38561.

www.ingramcontent.com/pod-product-compliance
Lightning Source LLC
Chambersburg PA
CBHW071738080526
44588CB00013B/2081